THE
SUPERSYMMETRIC WORLD
The Beginnings of the Theory

THE
SUPERSYMMETRIC WORLD
The Beginnings of the Theory

Edited by

G. Kane
The Randall Laboratory, University of Michigan, USA

M. Shifman
Theoretical Physics Institute, University of Minnesota, USA

World Scientific
Singapore • New Jersey • London • Hong Kong

Published by

World Scientific Publishing Co. Pte. Ltd.

P O Box 128, Farrer Road, Singapore 912805

USA office: Suite 1B, 1060 Main Street, River Edge, NJ 07661

UK office: 57 Shelton Street, Covent Garden, London WC2H 9HE

British Library Cataloguing-in-Publication Data
A catalogue record for this book is available from the British Library.

The editors and publisher would like to thank Springer-Verlag for their kind permission to include the contribution by D. V. Volkov.

Cover Illustration by Julia Shifman (1988).

THE SUPERSYMMETRIC WORLD

ISBN 981-02-4522-X
ISBN 981-02-4539-4 (pbk)

Printed in Singapore by World Scientific Printers

FOREWORD

G. L. KANE

Randall Laboratory, University of Michigan, Ann Arbor, MI 48109

M. SHIFMAN

Theoretical Physics Institute,
University of Minnesota, Minneapolis, MN 55455

"... One of the biggest adventures of all is the search for supersymmetry. Supersymmetry is the framework in which theoretical physicists have sought to answer some of the questions left open by the Standard Model of particle physics.

Supersymmetry, if it holds in nature, is part of the quantum structure of space and time. In everyday life, we measure space and time by numbers, "It is now three o'clock, the elevation is two hundred meters above sea level," and so on. Numbers are classical concepts, known to humans since long before Quantum Mechanics was developed in the early twentieth century. The discovery of Quantum Mechanics changed our understanding of almost everything in physics, but our basic way of thinking about space and time has not yet been affected.

Showing that nature is supersymmetric would change that, by revealing a quantum dimension of space and time, not measurable by ordinary numbers. .... Discovery of supersymmetry would be one of the real milestones in physics."

E. Witten[1]

The history of supersymmetry is exceptional. All other major conceptual developments in physics and science have occurred because scientists were trying to understand or study some established aspect of nature, or to solve some puzzle arising from data. The discovery of supersymmetry in the early 1970's, an invariance of the theory under interchange of fermions and bosons, was a purely intellectual achievement, driven by the logic of theoretical development rather than by the pressure of existing data. Thirty years elapsed from the time of discovery, immense theoretical effort was invested in this field, over 30,000 papers published. However, none of them can claim to report the experimental discovery of supersymmetry (although there are some hints, of which we will say more later). In this respect the phenomenon is rather unprecedented in the history of physics. Einstein's general relativity, the closest possible analogy one can give, was experimentally confirmed within several years after its creation. Only in one or two occasions have theoretical predictions of a comparable magnitude had to wait for experimental confirmation that long. For example, the neutrino had a time lag of 27 years.

It would not be an exaggeration to say that today supersymmetry dominates theoretical high energy physics. Many believe that it will play the same revolutionary role in the physics of the 21st century as special and general relativity did in the physics of the 20th century. This belief is based on aesthetical appeal, on indirect evidence, and on the fact that no theoretical alternative is in sight.

The discovery of supersymmetry presents a dramatic story dating back to the late 1960's and early '70's. For young people who entered high energy physics in the 1990's this is ancient history. Memories fade away as live participants of these events approach the retirement age; some of them have already retired and some, unfortunately, left this world. Collecting live testimonies of the pioneers, and preserving them for the future, seems timely given the impact supersymmetry has already produced on the development of particle physics. Having said that, we note that this book did not appear as a result of a conscious project. Both editors had collected some materials for other activities[2,3] and became aware of the other's interest and materials. Many people have been interested in how supersymmetry originated —the question often is asked in informal conversations—and how it can be such an active field even before direct experimental confirmation. We finally decided to combine materials, invite further ones, and edit this volume that makes available a significant amount of information about the origins of this intellectually exciting area. Most of it is in the words of the original participants.

In the historical explorations of scientific discoveries (especially, theoretical) it is always very difficult to draw a "red line" marking the true beginning, which would separate "before" and "after." Almost always there exists a chain of works which interpolates, more or less continuously, between the distant past and the present. Supersymmetry is no exception, the more so because it has multiple roots. It was observed as a world-sheet two-dimensional symmetry[a] in string theories around 1970; at approximately the same time Golfand and Likhtman found the superextension of the Poincaré algebra and constructed the first four-dimensional field theory with supersymmetry, (massive) quantum electrodynamics of spinors and scalars. Within a year Volkov and collaborators (independently) suggested nonlinear realizations of supersymmetry and started the foundations of supergravity. Using the terminology of the string practitioners one can say that the first supersymmetry revolution occurred in 1970-71 as the idea originated.[b] The second supersymmetry revolution came with the work of Wess and Zumino in 1973. Their discovery opened to the rest of the community the gates to the Superworld. The work on supersymmetry was tightly woven in the fabrique of contemporary theoretical physics. During the first few years of its development, there was essentially no interest in whether or how supersymmetry might be relevant to understanding nature and the traditional goals of physics. It was "a solution in search of a problem." Starting in the early 1980's, people began to realize that supersymmetry might indeed solve some basic problems of our world. This time may be characterized as the third supersymmetry revolution.

So, how far in the past one should go and where one should stop in the book devoted to the beginnings?

The above questions hardly have unambiguous answers. We decided to start from Ramond, Neveu, Schwarz, Gervais, and Sakita whose memoirs are collected in the chapter entitled *The Predecessors*, which opens the book. The work of these authors can be viewed as precursive to the discovery of supersymmetry in four dimensions. It

---

[a]The realization that the very same string theories gave rise to supersymmetry in the target space came much later.

[b]In the Marxist terminology it would be more exact to say that this was a prerevolutionary situation. This nuance is too subtle, however, and cannot be adequately discussed in this article.

paved the way to Wess and Zumino.

The central in the first part of the book is Chapter 2 presenting *The Discovery*. It contains recollections of Likhtman, Volkov, Akulov, Koretz-Golfand (Yuri Golfand's widow) and the 1999 Distinguished Technion Lecture of Prof. J. Wess, in which the basic stages of the theoretical construction are outlined.[c] Chapter 3 is devoted to the advent of supergravity. The fourth chapter is entitled *The Pioneers*. The definition of pioneers (i.e. those who made crucial contributions at the earliest stage) is quite ambiguous, as is the upper cut off in time which we set, *the summer of 1976*. By that time no more than a few dozen of original papers on supersymmetry had been published.

The selection of the contributors was a difficult task. We were unable to give floor to some theorists who were instrumental at the early stages (e.g. R. Arnowitt, L. Brink, R. Delbourgo, P.G.O. Freund, D.R.T. Jones, J.T. Łopuszański, P. Nath, Y. Ne'eman, V.I. Ogievetsky, A. Salam, E. Sokatchev, B. de Wit). Some are represented in other chapters (e.g. S. Ferrara whose 1994 Dirac Lecture is being published in Chap. 3.) Others are beyond reach. This refers to Abdus Salam and Victor Ogievetsky. The latter, by the way, wrote (together with L. Mezincescu) the first comprehensive review on supersymmetry which was published in 1975.[4] Even now it remains an excellent introduction to the subject, in spite of the 25 years that have elapsed.

The question of where to draw the line tortured us, and we bring our apologies to all the pioneers who "fell through the cracks."

The second part of the book is an attempt to present a historical perspective on the development of the subject. This task obviously belongs to the professional historians of science; the most far-sighted of them will undoubtedly turn their attention to supersymmetry soon. For the time being, however, to the best of our knowledge, there are no professional investigations on the issue. There was avail-

---

[c]Unfortunately, our (probably, awkward) attempts to convince Prof. B. Zumino in the usefulness of this book failed—we were unable to obtain his contribution.

able a treatise written by Rosanne Di Stefano in 1988 for a conference proceedings which were never published. This is a very thorough and insightful review. On the factual side it goes far beyond any other material on the history of supersymmetry one can find in the literature. There are some omissions, mostly regarding the Soviet contributors, which are naturally explained by the isolation of the Soviet community before the demise of the USSR and relative inaccessibility of several key papers written in Russian. The Yuri Golfand Memorial Volume[3] which contains the English translation of an important paper by Golfand and Likhtman[5] as well as a wealth of other relevant materials, fills the gap. In addition, Springer-Verlag has recently published Memorial Volumes in honor of Dmitry Volkov[6] and Victor Ogievetsky,[7] which acquaint the interested reader with their roles to a much fuller extent than previously.

The coverage of certain physics issues in Di Stefano's essay required comment; in a few cases we added explanatory footnotes. Di Stefano's essay is preceded by a relatively short article written by the late Prof. Marinov. It is entitled "Revealing the Path to the Superworld" and was originally intended for the Golfand Volume. This article presents "a bird's eye view" on the area. On the factual side it is much less comprehensive than Di Stefano's, but it carries a distinctive flavor of the testimony of an eye witness. Moreover, it reveals the mathematical roots of the discovery, an issue which is only marginally touched in Di Stefano's essay.

We are certainly not professional historians of science; still we undertook a little investigation of our own. Often students ask where the name "supersymmetry" came from? It seems that it was coined in the paper by Salam and Strathdee[8] where these authors constructed supersymmetric Yang-Mills theory. This paper was received by the editorial office on June 6, 1974, exactly eight months after that of Wess and Zumino. Super-symmetry (with a hyphen) is in the title, while in the body of the paper Salam and Strathdee use both, the old version of Wess and Zumino, "super-gauge symmetry," and the new one. An earlier paper of Ferrara and Zumino[9] (received by the editorial office[d] on May 27, 1974) where the same problem of super-Yang-Mills was addressed, mentions only supergauge invariance and supergauge transformations.

⋆ ⋆ ⋆

Supersymmetry is nearly thirty years old. It seems that now we are approaching the fourth supersymmetry revolution which will demonstrate its relevance to nature. Although not numerous, we do have hints that this is the case. They are: (a) supersymmetry allows a stable hierarchy between the weak scale and the shorter distance scales such as the Planck scale or unification scale, (b) supersymmetry provides a way to understand how the electroweak $SU(2) \times U(1)$ symmetry is broken, so long as the top quark came out heavy (which it did), (c) gauge couplings unify rather accurately when superpartners are included in the loops,[e] (d) the Higgs boson is predicted to be light (LEP gives $M_H < 200$ GeV), and (e) the lack of any deviations from Standard Model predictions in the precision data at LEP and in other experiments is consistent with supersymmetry (it was anticipated that these deviations would be invisible).

Certainly, at the moment the indications are not conclusive. However inconclusive, they are the source of hope and enthusiasm for phenomenologically oriented theorists and experimentalists who would like to keep high-energy physics in the realm of empirical science.

Another aspect which came to limelight recently is the fact that supersymmetry became instrumental in the solution of highly nontrivial dynamical issues in strongly coupled non-supersymmetric theories, which defied solutions for decades. That of course does not imply that nature is supersymmetric, but it does add to the interest in supersymmetry.

---

[d]The editorial note says it was received on May 27, 1973. This is certainly a misprint; otherwise, the event would be acausal.

[e]An alternative way to say this is to say that the value of the weak mixing angle at the weak scale can be calculated accurately if one sets it to the value predicted by a unified theory at the unification scale.

Summarizing, in this book we bring together contributions from many of the key players of the early days of supersymmetry. We leave its relevance to our world to a future project.

28 August 2000.

References

1. This quotation is from Witten's Foreword to G. Kane's book cited in Ref. 2.
2. G. Kane, *Supersymmetry: Unveiling the Ultimate Laws of Nature* (Perseus Books, 2000).
3. *The Many Faces of the Superworld*, Yuri Golfand Memorial Volume, Ed. M. Shifman (World Scientific, Singapore, 2000).
4. V.I. Ogievetsky, and L. Mezincescu, *Usp. Fiz. Nauk* **117**, 637 (1975), [*Sov. Phys. – Uspekhi* **18**, 960 (1975)].
5. Yu.A. Golfand and E.P. Likhtman, *On the Extensions of the Algebra of the Generators of the Poincaré Group by the Bispinor Generators*, in I. E. Tamm Memorial Volume *Problems of Theoretical Physics*, Eds. V.L. Ginzburg *et al.*, (Nauka, Moscow, 1972), page 37 [English translation: Ref. 3, page 45].
6. *Supersymmetry and Quantum Field Theory*, Proceedings of the D. Volkov Memorial Seminar, Kharkov, Ukraine, January 1997 [Lecture Notes in Physics, Vol. 509] (Springer-Verlag, 1998).
7. *Supersymmetries and Quantum Symmetries*, Proceedings of the International Seminar on Supersymmetries and Quantum Symmetries Dedicated to the Memory of V.I. Ogievetsky, Dubna, Russia, July 1997 [Lecture Notes in Physics, Vol. 524] (Springer-Verlag, 1999).
8. A. Salam and J. Strathdee, "Super-symmetry and non-Abelian gauges", *Phys. Lett.* B **51**, 353 (1974).
9. S. Ferrara and B. Zumino, "Supergauge invariant Yang-Mills theories", *Nucl. Phys.* B **79**, 413 (1974).

General acknowledgments

The editors would like to express their sincere gratitude to all participants of the project. Special thanks go to Alex Roitman who was responsible for typesetting this volume in LaTeX, and to Eric Barrett for expert advice on the English Grammar.

We are grateful to Julia Shifman for allowing us to use one of her early watercolors for the cover illustration and to M. Fitzer and T. Shifman for making available to us fragments of an interview they conducted with John Schwarz in November 1999.

List of participants

V. Akulov, R. Di Stefano, P. Fayet, S. Ferrara, G.-L. Gervais, G.L. Kane, N. Koretz-Golfand, E. Likhtman, M. Marinov, A. Neveu, L. O'Raifeartaigh, P. Ramond, B. Sakita, J. Schwarz, M. Shifman, M. Sohnius, V. Soroka, J. Strathdee, D. Volkov, J. Wess, P. West.

CONTENTS

Local Supersymmetry (Supergravity)

The Pioneers

The Historical Perspective

1

The Predecessors

P. Ramond

A. Neveu

J. Schwarz

J.-L. Gervais

B. Sakita

EARLY SUPERSYMMETRY ON THE PRAIRIE

P. RAMOND

Physics Department, University of Florida,
Gainesville, FL 32611

My first postdoc started at NAL in September 1969, one of a group of five postdocs, with no senior theorists. NAL, which was to become the Fermi National Accelerator Laboratory, was being energetically built under the guidance of Bob Wilson. In the fall, I was introduced to Professor Nambu by a fellow postdoc and Chicago graduate Lou Clavelli. He was very kind to us, expressing interest in our work on the group-theoretical construction of the dual amplitudes, and even inviting us to lunch at the Quadrangle Club! This was a welcome contrast to visiting theorists who showed little or no interest in our science, with some exception (see later).

Earlier that summer, armed with a fresh Ph.D., I had been introduced to the Veneziano model by working in Trieste on deriving the triple Reggeon vertex with J. Nuyts and H. Sugawara. We published nothing, as our work, which was cast in an awkward notation, was superseded by Sciuto's paper using the oscillator formalism. Still, I felt attracted to the model by its elegance and because I could understand some of it, and remained determined to work on it.

One day, Bob Wilson came to our office and said: "All theorists must go to Aspen this summer." As employees we had no choice, and all of us went there in the summer of 1970. This is where my story begins.

I found the Aspen Center for Physics to be just the right kind of place to wrench me away from the rather technical calculations Lou and I had been involved in: I started thinking about the Veneziano model in a different way in terms of its relationship to point-particle dynamics. For instance, the Klein-Gordon operator in momentum space $(p^2 - m^2)\phi = 0$, could be generalized to the Virasoro operator $L_0 - 1$, which could also be written as the square of a "momentum" operator

$$P_\mu(\tau) = \frac{dQ_\mu(\tau)}{dt}$$

averaged over the inner periodic variable τ, where $Q_\mu(\tau)$ was the generalized position introduced by Fubini and Veneziano. Hence, I thought that the transition from point particle to the Veneziano model was simply the correspondence

$$p_\mu \rightarrow P_\mu(\tau) \,,$$

such that the point-particle momentum and positions are given by

$$p_\mu = \langle P_\mu(\tau) \rangle_0 \,, \qquad x_\mu = \langle Q_\mu(\tau) \rangle_0 \,,$$

where the average simply meant integration over τ. In that spirit, the Klein-Gordon operator generalizes to

$$p^2 \rightarrow P_\mu(\tau)P^\mu(\tau) \,,$$

which depends on the internal variable. The Klein-Gordon equation then becomes the average of the product, yielding the simple equation of motion

$$\langle P_\mu(\tau)P^\mu(\tau) \rangle_0 = m^2 \,.$$

The other moments (the operators introduced by Virasoro) form the algebra

$$[L_m, L_n] = (n - m)L_{m+n} \,.$$

(Of course, I did not know about the c-number, found by the late J. Weiss). As they all started like $p_\mu a_n^{\dagger\mu}$, they played the role of decoupling the ghost states, although I could not formulate this statement in concrete terms. I found this correspondence strikingly simple, and I checked that it also worked for the Lorentz generators

$$x_\mu p_\nu - x_\nu p_\mu \rightarrow \langle Q_\mu(\tau)P_\nu(\tau) - Q_\nu(\tau)P_\mu(\tau) \rangle_0 \,.$$

It must have been right at the end of my Aspen stay when I thought how to apply this correspondence principle, but I felt in no hurry, especially given the intellectual pleasure one gets from leisurely playing with an intriguing idea.

Upon my return from Aspen — its atmosphere I enjoy to this day, with eternal thanks to Bob Wilson — I wrote up the correspondence principle for the bosonic case in a paper (NAL Report THY 7), and submitted it to Physics Letters B in October 1970. It was promptly rejected, and I withdrew it from consideration (many years later Maurice Jacob the editor of the journal told me he was surprised when I decided to withdraw it, but I did not know the rules of the game!)

Although pretty dejected, I turned my full attention to apply the newly-found correspondence principle to the Dirac equation

$$(\gamma_\mu \, p^\mu - m)\Psi = 0 \ .$$

Lillian had just gone to Princeton on a training course for her engineering job, and I was intent on getting to the bottom of this problem before her return. I remember working with almost no interruptions for a period of several weeks.

Back to Physics. It was obvious how to proceed: invent a generalized gamma matrix $\Gamma_\mu(\tau)$ with the properties

$$\gamma_\mu = \langle \Gamma_\mu(\tau) \rangle_0 \ ,$$

and the anticommutation relations

$$\{\Gamma_\mu(\tau), \ \Gamma_\nu(\tau')\} = \delta(\tau - \tau') \ .$$

This implied

$$\Gamma_\mu(\tau) = \gamma_\mu + \gamma_5 \sum_{n=1} (b_\mu^{n\dagger} e^{in\omega\tau} + b_\mu^n e^{-in\omega\tau}) \ ,$$

where $\{b_\mu^{n\dagger}, \ b_\nu^m\} = g_{\mu\nu}\delta^{mn}$ are the Fermi oscillators, and yet transform as Lorentz vectors, an association that could have seemed puzzling (it was not until much later that the magical properties of ten dimensions threw some light on this). The generalized Dirac operator is now obvious

$$\gamma_\mu \, p^\mu \ \rightarrow \ \langle \Gamma_\mu(\tau) P_\mu(\tau) \rangle_0 \equiv F_0 \ .$$

The decoupling operators are simply the other moments

$$F_n \equiv \langle \Gamma_\mu(\tau) P_\mu(\tau) \rangle_n = \langle e^{in\omega\tau} \Gamma_\mu(\tau) P_\mu(\tau) \rangle_0 \ .$$

The rest was simple algebra, which generated a new realization of the Virasoro algebra as well as a curious algebra which had both commutators and anticommutators,

$$\{F_n, \ F_m\} = 2L^F_{m+n} \ ; \qquad [L_n, \ L_m] = \omega(m - n)L_{m+n} \ ,$$

where now the new Virasoro operator is the sum of two terms

$$L_m = \langle e^{in\omega\tau} P_\mu(\tau) P^\mu(\tau) \rangle - \frac{i}{4} \langle e^{in\omega\tau} \Gamma_\mu(\tau) \frac{d\Gamma^\mu(\tau)}{d\tau} \rangle \ .$$

It was clear that F_n were new decoupling operators associated with the Fermi oscillators. Also, the Lorentz generators worked out as the spin part generalized correctly through the correspondence principle

$$[\gamma_\mu, \gamma_\nu] \ \rightarrow \ \langle [\Gamma_\mu(\tau), \Gamma_\nu(\tau)] \rangle_0 \ .$$

I worked for a period of several weeks almost with no interruption, and with increased excitement as I performed all kinds of checks, which worked. It was clear to me that I had generalized Dirac's equation to the Veneziano model.

I then climbed in my little sports car to see Lillian at Princeton, and also my mentor, Professor Nambu, who was spending the fall at the Institute for Advanced Studies (where I met Mike Green for the first time). Nambu's reaction to the work was very positive. Back at NAL, I proceeded to write up the work, although I continued to play with it, realizing that the point particle can be viewed as a limit of the Veneziano case by varying the slope parameter and that the zero point energy of the boson and fermion operators cancelled. I procrastinated until the Holiday season. Just before Christmas 1970, I sent the fermion paper (NAL THY-8) to the Physical Review, and the (previously rejected) boson paper to Il Nuovo Cimento. Both papers were accepted for publication.[1,2] Although the paper fell short of an interacting theory, I found myself unable to concentrate for I was exhausted (and fell ill and bedridden a bit later) and had to find a job: although we had been hired with the promise of longer term appointments (because of the lack of senior theorists), Wilson abruptly decided to terminate us after two years, and we had to find a job. As a result, I did not go back to the interacting case until many months later.

Several anecdotes around that time come to mind. I explained the fermion work to my colleague Don Weingarten (at a luncheonette in downtown Wheaton!), and I remember his answer for he said I was "set for life"!

According to Bob Wilson, the role of the theorists at the laboratory is to entertain visiting theorists, and almost every week, some luminary came to visit. The schedule left little time for discussion, as it involved a climb up a converted silo near the main ring then under construction, a free lunch at a local restaurant, and a tour of the Village. One such visitor that fall was Stanley Mandelstam. In the morning, I managed to tell Mandelstam about the Dirac equation work; he listened carefully but gave no indication of interest, nor disinterest. Later that afternoon, I asked him what he thought. His answer startled me for he said: "You ask me for an answer right now, but I have to study this further. You claim to have solved a problem that many people including my colleagues at Berkeley have been trying to solve. I do not know who you are, and from what you told me, I cannot tell if you have succeeded, but I will study it and let you know." I was very startled and frankly depressed as the equation was by now so obvious to me, but then again he did not know all the details that I knew. Rarely have I heard such an honest and forthright answer to a question. Stanley was completely matter-of-fact and honest. As he promised, he studied it, quickly understood more than I did, and made every effort to bring the work to people's attention. I owe him a great deal.

In January 1971, Lou Clavelli and I went to the auction block, the annual meeting of the APS then in New York City to look for a job. Lou asked me to tag along to New Haven for a few days, where he had been a postdoc and married. He was staying with his in-laws and kindly thought I could visit Yale. This was a great visit for I met my lifelong friend the late Dick Slansky, and also gave an impromptu talk on the Dirac equation. The following Monday, back at NAL, I received a call from Charlie Sommerfield offering me a job (actually half a job as an instructor, but never mind, at least a job). Right away I called up Lillian all excited, and then Charlie a few minutes later to accept. Years later Charlie expressed amazement at the prompt acceptance, but I had never before (or since) been offered a job that fast either!

References

1. P. Ramond, "Dual Theory for Fermions," *Phys. Rev.* **D3**, 2415 (1971).
2. P. Ramond, "An Interpretation of Dual Theories," *Nuov. Cim.* **A4**, 544 (1971).

THIRTY YEARS AGO

ANDRÉ NEVEU

Laboratoire de Physique Mathématique,
CNRS–Université Montpellier II,
34095 Montpellier, France

The origins of my involvement in the superstring saga can be traced back to the rather fortuitous and amusing circumstances of my first physical encounter with Pierre Ramond.

Actually, in 1968-1969, while I was preparing my "thèse de troisième cycle" (an analogue of the Ph.D.) in Orsay, I had already heard (in very positive terms) about Pierre from Jean Nuyts (then in Orsay), with whom he had seemingly signed the papers (without having met, if I remember) on crossing-symmetric partial wave amplitudes. Pierre's Ph.D. in Syracuse was devoted to crossing symmetry on which he worked with A. P. Balachandran, his advisor.

In September 1969, I had just finished my four year curriculum at the École Normale Supérieure in Paris. I had been awarded the "Jane Eliza Procter Fellowship" (endowed by Procter and Gamble) which Princeton University reserves each year for an alumnus of the E.N.S.; with it came a Fulbright travel grant which enabled me to choose the ship "France" for my first trans-Atlantic crossing. The ship had a small and pleasant library with a few desks. I was spending many hours there, studying in detail the latest preprints on dual resonance models, as they were called. One afternoon, leaving all my materials spread all over the desk, I walked out of the library, called by an urgent need... During these two minutes when I was absent, Pierre Ramond walked in and looked around for a vacant desk. There appeared to be only one – mine. He walked up to it, realized that it was not really vacant, but was shocked to see on it the Fubini-Veneziano paper on the factorization of dual resonance models, the very same paper he was studying at that moment! He quickly went back to his cabin to make sure that what he had just seen was not his copy! Reassured about his sanity, he came back to the library, wondering who could be the fellow interested in such an esoteric topic. By this time I, too, was back. You can imagine easily the next few hours. This is how we became friends. After spending his summer vacation in France, Pierre was on his way to the National Accelerator Laboratory (now called Fermilab) for his first postdoc.

At the Physics Department of Princeton University, there was no preprint library. No need was felt for it since all the important people there received important preprints themselves, and then disseminated the news which might be of any use. Would a short and partly speculative paper by a still relatively obscure postdoc at Fermilab have been considered important? And reach me? Perhaps, but probably not soon enough. As I realized during a visit to Berkeley a few weeks later, the next breakthrough would very likely have been made there, and probably within a couple of months, by one or several of the following: Korkut Bardakçi, Martin Halpern, Michio Kaku, Stanley Mandelstam, Charles Thorn, and ... Miguel Virasoro, who was a postdoc there at that time. But thanks to the crossing on the "France," Pierre had sent John Schwarz and me his preprint on the free Ramond model. We immediately realized that there was something very deep there, and proceeded to build an interacting theory of one fermion line with mesons. By factorizing in the fermion-antifermion channel, we discovered that the mesonic trajectories had many interesting properties, and decided to build a purely bosonic sector embedding what is now called the superconformal algebra. I must confess I was not very comfortable with the complexities of the γ matrices of the fermion sector (for extenuation, I plead youth), and, moreover, what was to be expected of the duality properties of the fermionic amplitudes was far from clear at that time.

So, as for me, there was some amount of luck in our discovery. There was also a certain degree of unconsciousness, as still transpires through a careful analysis of our first paper on the Neveu-Schwarz sector. As far as I remember, I was not even really conscious of the fact that introducing modes with a space-time vector index meant introducing such a huge number of ghosts that one had better first sit back and think of a new symmetry which would give a chance of eliminating them, instead of forging ahead as we did, computing amplitudes and residues almost at random, and realizing that there was, indeed, a miracle. The Ramond superconformal algebra had been our guide in the construction of a few bosonic amplitudes, by factorizing off a free fermion line, but in the Neveu-Schwarz sector, the subtleties of the F_1 and F_2 formalisms could not be guessed in advance, so sitting back and thinking would probably have led us nowhere. I only remember that I was caught by the mathematical beauty of the superconformal algebra, and just hoped, more or less consciously, that, one way or another, in connection with it, a ghost cancelling miracle would occur...

This is an instance where I may quote the French proverb *"Aux innocents les mains pleines"* roughly translated "Hands full for the innocent," remarking that *innocent* in French generally means not just the opposite to guilty, but, in this case, rather refers to the innocence of youth if one is kind, or to feeblemindedness if one is not (the most usual occurrence).

STRINGS AND THE ADVENT
OF SUPERSYMMETRY:
THE VIEW FROM PASADENA [a]

JOHN SCHWARZ

Physics Department, California Institute of Technology, Pasadena
CA 91125

Q. To warm up I would like to ask how your early work was related to supersymmetry.

A. The earliest version of string theory (developed in the late 1960's to describe hadron interactions) suffered from various unphysical features. In particular, the spectrum contained a tachyon but no fermions. This motivated the search for a more realistic string theory. The first significant success was made in January 1971 by Pierre Ramond[1] who constructed a string analog of the Dirac equation. At about the same time, Andre Neveu and I were in Princeton constructing a new bosonic string theory. Neveu and I quickly realized[2] that the two constructions were different facets of a single theory and (along with Charles Thorn) we constructed[3] an interacting string theory containing our bosons and Ramond's fermions. In 1972 I showed that consistency of the theory requires that the space-time dimension is 10 and that the ground state fermions are massless.

The Ramond-Neveu-Schwarz string theory actually contains local space-time supersymmetry, but we were slow to realize that. The realization came after supersymmetry in four dimensions was thoroughly studied by others.

Q. What can you say about the string theory today?

A. My impression of the current state of the string theory is that we've now got overwhelming evidence that there is a unique fundamental theory that contains gravity and other forces. But that we only understand this theory in certain special limits, and we don't

---

[a]These excerpts are based on the interview conducted by Melitta Fitzer and T. Shifman, Santa Barbara, November 19, 1999 (the transcript was prepared by A. Roitman), and on a private letter from J. Schwarz to G.L. Kane.

have a complete understanding of it. It has not been fully formulated. And that until it is fully formulated, that there are many important questions that we are unable to answer. So, while we've made enormous progress in seeing the relations between the limiting cases which we used to think of as different theories, we now realize they are actually different limits of a single theory. That it looks like it is all coming together. But there are really major questions about how we are going to relate this theory to the real world. These questions are partly of a more practical character, and partly of a more theoretical character. There are big qualitative questions that we do not know how they are going to work out. But I think everyone who works on this field is so taken by the beauty of the theory that they have little doubt that the right answer is there to be found, though we do not know exactly what it is. For example, one of the big questions that we do not understand, is what determines how much energy there is associated with completely empty space. This is something referred to as a cosmological constant. From observations we know that this number is extremely small, maybe zero. Yet at a theoretical level we don't see any reason why it should be small. Until we understand that, it's clear that there are big qualitative questions that remain to be explained. Supersymmetry, if it were exact, could explain the vanishing of the cosmological constant. There's been a lot of progress in explaining other qualitative questions about the relationship between strings theory and different quantum field theories, how to learn about the behavior in these theories when the interactions become very strong. This is usually very difficult, but in situations where there is supersymmetry, which is a very special kind of symmetry, one can make some fairly precise statements.

Q. I understand that supersymmetry is a very unusual symmetry. Could you, please, explain why is it so special?

A. Yes, supersymmetry is a very interesting kind of symmetry that has developed within the last 30 years or so. There are different ways of explaining it. It can be thought of as an extension of the symmetry that would be associated with space and time, to some more abstract directions. There are two classes of particles in quantum mechanics, which go by the names of bosons and fermions. Ordinary, more familiar kinds of symmetries – like rotational symmetry of a sphere or something like that – only relate different bosons to one another and different fermions to one another. But supersymmetry can relate these two different classes of particles, the bosons to the fermions. And that is of rather deep significance, because, roughly speaking, you can think of bosons as the particles responsible for forces in nature, and fermions as the particles that are responsible for stable matter. So if you can find some principle to relate these two types of particles then that is quite profound. Maybe, the first experimental evidence for the string theory might arise from the discovery of supersymmetry. The discovery of supersymmetry would not prove that the string theory was right, and its non-discovery would not prove that it was wrong. But, certainly, the discovery would be enlightening and informative, and I think it would be very encouraging for those people who are pursuing string theory.

Q. Thirty years ago, when you started, you were fairly alone in this field, right?

A. Well, I would not put my lonely period back this far. The subject of string theory arose about 30 years ago. It was extremely popular for about five years, there were hundreds of people working on it. And they were working on it for the specific problem of describing strong nuclear forces. And then, around 1973-74, a better theory for the strong nuclear forces arose. It now is referred to as QCD. Everyone became convinced that that was right. String theory was failing in many details to provide a satisfactory theory. So, at that point almost everyone stopped working on the string theory. And I was one of the few who kept working on it despite this failure, just because we were so taken by the mathematical beauty of the subject. We felt that it should be useful for something. One of the problems that we were having with the theory was that it contained a particular kind of particle that did not belong to the strong interactions. Eventually, in 1974, Joël Scherk and I realized that these particles had exactly the right properties to describe gravity.[4] We said: "Let's forget about strong interactions, let's use this theory for completely different purpose than for which it was originally invented, and use it as a theory of gravity. And then other forces would follow."

So my focus completely shifted, almost overnight, at some point in 1974, when we realized that this was the basis for providing a theory of gravity that would be compatible with quantum theory. More traditional approaches to adding gravity into quantum field theory ran into various difficulties. It was clear to us that in the string theory framework those difficulties would be avoided. So the period of loneliness that you refer to was really the subsequent 10 years, from 1974 to 1984, when I was pursuing that program with a number of collaborators, most notably Michael Green. Unfortunately, Joël Scherk died in 1980. I started working with Michael Green just about that time, and Lars Brink collaborated with us on a number of occasions as well. So we were just a handful of people pursuing this program. I was very enthusiastic about it. We were not completely ignored, we spoke at many conferences and so on, but most people were doing other things. Some rather related, but not exactly string theory. Others were working on supersymmetry, supergravity, Kaluza-Klein theory, all of which are parts of what is now considered string theory. But the string element was missing in the work that the others were doing. In any case, there were a number of developments in 1984, one of which Michael Green and I were responsible for, which led to the subject suddenly becoming very popular. And since then it's been a very active subject. It's had its ups and downs since then. There was another flurry of discovery in 1994, ten years later, which further boosted the community into the general acceptance of the subject. By now it is considered pretty mainstream. All the major physics departments are trying to hire new faculty to work in this area. So that's quite a change from the way things used to be, when famous physicists made some rather derogatory statements about it.

Q. When you pursue an idea, when you know within yourself that something feels interesting or right, and there are people around who do not agree with you, how do you deal with this?

A. I've never felt that there's much of a benefit of arguing, at length, with people about what was right and what was wrong. I present my ideas, and they can accept it or reject it, as they wish, I would leave it at that.

In a field like ours, eventually, it seems to me, the power of the idea will carry the day, although it may take some time. I prefer to let the ideas speak for themselves.

Q. I understand that to get grants and promotions the physicists here, in America, have to publish papers which are cited by other researchers. If you work in an area that nobody else is in, how do you solve this problem?

A. Well, I've never had problems of this type, even in the period when there was not much interest in what I was doing. I was employed at Caltech. Murray Gell-Mann, who was an outstanding scientist, was in our group in Caltech. He was responsible for bringing me there. He felt that what I was doing might have some possibility of being right and deserved to be supported. So I had a pretty good situation all along. I didn't have to struggle to get grants.

Q. But emotionally, did you feel that you were separated from others?

A. I tend to forget this with the passage of time. But if I think back, yes, there's no doubt that there were periods in which I felt very frustrated that there wasn't more acceptance of what I was doing. There's a funny thing. When the subject became popular, around 1984, I felt at that time that since I've been working on it for so many years, I had a big advantage over all these people who jumped in. And I would be able to develop many important ideas, more rapidly than they were able to, because I had all this experience. But that turned out to be completely false, and within months I was completely overtaken by all these newcomers.

Q. Sometimes people refer to the string theory as to a theory of everything. Any comments?

A. A theory of everything! I've been quoted on other occasions as being very negative about it, and I am. The phrase was introduced by others, and I think it was very unfortunate for us. In a very limited sense, what we are trying to do is fundamental to everything else, and that is what I guess this phrase was meant to represent. But it overlooks something that is terribly important. Even though, in principle, when you have the most microscopic rules, that determines how the things work on very short scales, as you go to larger scales, there are qualitatively new phenomena that arise, that are almost impossible to understand in terms of the underlying microscopic theory. By knowing string theory, we are not going to make any progress at all in understanding these other subjects. For instance, if you ask about human activities, life, I am certain that a complete understanding of string theory is not going to add any light on that. It is really not what we are addressing.

There is an analogy in the past to that. Around 1930 or so, when quantum mechanics was well understood, there was a famous statement due to Paul Dirac. He said that now all of chemistry is solved, in principle. In the same sense he was correct, the basic equations at that point were known. But getting from there to what the chemists actually observe involves computations that are so complex that they are completely out of the question, even for future computers that might exist in a hundred years from now. What better computers will allow you to do, is to talk about molecules that involve three atoms rather than two, or four atoms rather than three. But to understand macroscopic properties of matter based on understanding these microscopic laws is just unrealistic. And even though the microscopic laws are, in a certain strict sense, controlling what happens at the larger scale, they are not the right way to understand that. And that is why this phrase, "theory of everything," sounds sleazy.

Q. What if it is true that there are extra dimensions of a large size, of the type people discuss since 1996?

A. There are people that are speculating that extra dimensions are much larger than has traditionally been assumed to be the case by people, including myself, who work on this subject. My guess is that it is unlikely that this is the case. I would be quite surprised if the extra dimensions turned out to be as large as these other people suggest. Well, if they are right – strictly speaking, it is not excluded, I mean, experimentally – that would be very exciting, because that would mean that things could actually be studied experimentally. And, maybe, they would be useful for something, I don't know.

Q. You mentioned the beauty of the string theory. Could you elaborate a little bit?

A. Different people see it in different ways. A standard way of explaining this is that you have a rather simple-looking equation that explains a wide range of phenomena. That is very satisfying. Famous examples of that are due to Newton, Einstein, and so forth. We do not actually have a concise formula in string theory that explains a lot of things, so the beauty is of a somewhat different character here. What, I think, captures people in the subject is when they discover that they are dealing with a very tight mathematical system that incorporates things that nobody has understood yet. And then they find a part of that. So there is a deeper structure than we have understood. When you do some complicated calculation, and then you discover that the answer is surprisingly simple, much simpler than you would have expected... And it sort of appears as if there were some sort of a mathematical miracle taking place. Of course, there are no miracles, and when you find something that looks like a miracle, that just means that there is some important concept that you haven't understood yet. When you experience that a few times, you really get seduced by the subject. That's been my experience, and I think of the other people who are now just as enthusiastic, and, maybe, even more so, than I am, have had similar experiences.

Q. Is this a flash of the insight that keeps you in physics? What made you choose physics, to begin with, and what keeps you there?

A. I feel extremely fortunate that I am doing throughout my career exactly what I wanted to do. Nobody has ever tried to make me do anything else. So that's a fantastic situation to be in. And to be lucky enough to just be in the right place at a right time... It is very exciting.

Q. Do you think you ever in your life had a moment when you thought "I'm just throwing it to a dump, and I am going into a different business?"

A. No, I never had that. One thing that occurred to me, at the time when there wasn't interest in what we were doing, but I felt it was important, was that if I tried to look at my own situation very objectively, as an outsider, I don't know how anyone could decide whether or not I was a crackpot. Because, I think, you could find examples of people whose ideas did not hold up as well, that were as convinced and as dedicated to their work as I was. But I've never felt that I was a crackpot...

References

1. P. Ramond, *Dual Theory for Free Fermions*, Phys. Rev. **D3**, 2415–2418 (1971).
2. A. Neveu and J.H. Schwarz, *Factorizable Dual Model of Pions*, Nucl. Phys. **B31**, 86–112 (1971).
3. A. Neveu, J.H. Schwarz, and C.B. Thorn, *Reformulation of the Dual Pion Model*, Phys. Lett. **35B**, 529–533 (1971).
4. J. Scherk and J. H. Schwarz, *Dual Models For Nonhadrons*, Nucl. Phys. **B81**, 118–144 (1974).

REMEMBERING THE EARLY TIMES OF SUPERSYMMETRY[a]

JEAN-LOUP GERVAIS

Laboratoire de Physique Théorique de l'École Normale Supérieure
24 rue Lhomond, 75231 Paris CÉDEX 05, France[b]

As is well known, supersymmetry started long before the Iron Curtain was dismantled, and thus came into existence separately in the West and in the former Soviet Union. In order to speak about what I know personally, I will recall that in the West, superalgebras were first considered by A. Neveu, J. Schwarz and P. Ramond as a basic tool to eliminate negative norm states from the spinning string theories. When Neveu, Schwarz and Ramond first introduced them they were initially referred to as supergauges. These authors used the covariant harmonic oscillator approach, the only known technique at the time, without field-theoretic interpretation. Preparing the present text brought back wonderful memories of the time when my long lasting collaboration and friendship with Bunji Sakita[c] began. In 1970-71 we started to develop the world–sheet interpretation of the spinning string (unknown at that early time), extending earlier discussions of the purely bosonic case begun by H. Hsue B. Sakita and M. Virasoro.[1] We recognized that the Neveu-Schwarz-Ramond (NSR) models included world-sheet two-dimensional Dirac spinor fields, in addition to the world sheet scalar fields, common with the Virasoro model. We showed that the supergauges of the NSR models corresponded to the fact that the two-dimensional world-sheet Lagrangian was invariant under transformations with anticommuting parameters which mixed the scalar and spinor fields. This gave the first example of a supersymmetric local Lagrangian (albeit, two-dimensional).

---

[a]This is a slightly revised excerpt from a paper published in Yuri Golfand Memorial Volume *The Many Faces of the Superworld*, (World Scientific, Singapore, 2000).

[b]UMR 8549: Unité Mixte du Centre National de la Recherche Scientifique, et de l'École Normale Supérieure

[c]Then a Visiting Professor both at the Institut des Hautes Etudes, in Bures-sur-Yvette, and at the Laboratoire de Physique Théorique et Hautes Energies, University of Orsay, France.

This supersymmetry was closing only on shell for lack of auxiliary fields, and was two-dimensional. In the West, the problems of proceeding off shell and to four dimensions were initially solved by Wess and Zumino, as is well known.

These are the historical facts simply stated. It is worth trying to give a more personal picture of the birth of supersymmetry. Sakita himself has already given his own view.[2] As for me, the story really begins during the year 1968-69 which was a sort of a turning point. I had just returned to France after spending two years as a postdoc at New York University (NYU), where I had met J. Wess (a visitor for one year), B. Zumino (then the head of the Theory Group located at the Courant Institute), K. Symanzik, W. Zimmermann (at the time permanent faculty members) and D. Zwanziger (who was about to become a permanent member). Before that, my interest dealt mostly with dispersion relations, Regge poles, and S-matrix theory, but at NYU, I was fully converted to local field theory, and was much impressed by the power of symmetries in that context, be they local or global. Of course, that year saw the beginning of string theory which was, however, initially developed using the covariant operator method within the context of S-matrix theory, giving what looked like a realization of G. Chew's program. Regardless, local field theory also made wonderful progress on its own. The main problems of that time were the Adler-Bell-Jackiw anomaly, the spontaneous breaking of symmetries and the quantization of Yang-Mills theory. For the latter, the work of L. Faddeev and V. Popov was gradually becoming more and more popular. It is hardly necessary to say that these topics now belong to textbooks. At that time, the French Government was very generous with temporary positions, and a handful of key visitors came for long visits[d] during that wonderful year, including D. Amati, the late Benjamin Lee, T. Veltman,[e] and B. Zumino. I drew much inspiration from the very stimulating atmosphere they created, together with other permanent members. In particular, together with D. Amati and C. Bouchiat, I devised[3] the now standard method to compute loops in string theories, using coherent states,

---

[d]At the Laboratoire de Physique Théorique et Hautes Energies of Orsay (France) where I was working permanently at the time.

[e]Who had already undertaken to renormalize Yang-Mills theories.

and with B. Lee I showed[4] how to correctly quantize the linear σ model in the phase where the spontaneous symmetry breaking takes place.[f]

This is all to say that, when I first met Bunji Sakita in the fall of 1970, I was fully motivated to apply the field theory technique to string theories. Moreover, the year before I had shown[5] that the integrand of the Veneziano model was equal to the vacuum expectations of a product of scalar fields which were local functions of the Koba-Nielsen variables. This result, similar to an independent and better known work[6] by S. Fubini and G. Veneziano, was indeed a strong hint of the world-sheet field theory aspect of string theory. This viewpoint is now a common place, but at the time it was not at all popular among string theorists. A large majority preferred the operator method, which had achieved striking technical success.

Before we met, Sakita and his collaborators had already made important progress in developing world-sheet field theory technique using path integrals. On the one hand, H. Hsue, B. Sakita, and M. Virasoro[1] had shown how the analog model of H.B. Nielsen could be derived from the path integral over a free scalar two-dimensional field. On the other hand, Sakita had come with a draft of an article where he started to discuss Feynman-like rules for the Veneziano model using the factorization of path integrals over sliced Riemann surfaces. There were many basic problems left, and at the beginning, we spent a lot of time establishing a general scheme. This complicated work was not so well-received, although it contained many precursive results. The most important point was that we essentially relied on the conformal invariance of the path integral representation over scalar free fields in two dimensions.[7] Although we did not really consider the gauge-fixing problem at that time, we were pretty much convinced that conformal invariance of the free world-sheet action is at the origin of the negative norm state elimination. In the spring when we came across the first article of A. Neveu and J. Schwarz, we were motivated[8] to systematically discuss conformal field theories in two dimensions, as a way to classify string theo-

---

[f]It seems that G. 't Hooft —then a student at Cargese during the following summer— drew much inspiration for quantizing massive Yang-Mills theory, from B. Lee's lecture on spontaneous symmetry breaking.

ries, by defining what we called "irreducible fields" (now known as primary fields, following A. Belavin, A. Polyakov, and A.B. Zamolodchikov[9]). Considering only quadratic actions we recognized that only spin-zero and spin-one-half fields were possible, covering all existing critical string models of today. This was all well and good, except for one fact: the Neveu-Schwarz model had more ghosts and needed an additional negative-norm-state killing mechanism as compared with the Veneziano model. This motivated these authors, as well as Ramond in his seminal work, to introduce in the operator formalism a set of operators whose anticommutators gave the Virasoro generators. Sakita's visit in France was about to end, bringing with it the well-known burden of moving with a family. The coming deadline stimulated us a great deal, nevertheless. I quickly started to look for the possible symmetries of the action that would be the origin of the additional ghost killing. From the form of the NSR generators it was immediately clear that the transformation of the boson had to be a fermion and vice versa. It was not difficult to envisage that the action could be invariant, except for the mixing between commuting and anticommuting fields, which made everything very confusing. After many hesitations, Sakita and I solved the problem by introducing symmetry transformations with anticommuting parameters, out of which the supersymmetry of our NSR world-sheet action followed very simply. The paper was completed,[10] and, thus, for us, (conformal) supersymmetry was born, on July 22, 1971, just the day before Sakita departed from France to fill his new prestigious position at City College of New York.

In those days, anticommuting c-numbers were far from being understood. They were essentially a formal tool to derive perturbation expansions over fermionic fields from path integrals. This was how we used them in our second article,[8] where we derived the tree scattering amplitudes of the Neveu-Schwarz model from world-sheet perturbation. Considering symmetry transformations with anticommuting parameters was, for us, quite a step which we took rather reluctantly. Quite some time passed before our idea was taken seriously —even by us.

In the early seventies, the world-sheet dynamical approach to string theory was not well appreciated in our community, and our work did not get much attention in general. In December of 1971, Sakita delivered a talk about it at the Conference on Functional Methods in Field Theory and Statistics at the Lebedev Institute in Moscow, organized by E. Fradkin. On the way back, he stopped over in Paris. We wrote a summary of our work for the proceedings, which was sent to the organizers and circulated as a preprint — the complete proceedings themselves were never published.[9] In this way there was some early communication of our work to the Soviet scientific community where apparently it was well received.[2]

Scientifically our separation was at a very unfortunate moment for our research program. At that time, of course, there was no email. Telephones were expensive, and airmail slow. Moreover, Sakita had to deal with his new life and responsibilities, which he did rather successfully. We nevertheless continued our collaboration and met in person as often as possible. I was happy to visit him and his group in New York. I did not push very hard further in the direction of supersymmetry, however, to my regret. Other problems seemed more pressing. In the meantime the Nambu-Goto action and the Goddard-Goldstone-Rebbi and Thorn light-cone quantization had come out. Sakita and I[h] showed[12] how the latter may be recovered using path integral with the former action using the Faddeev-Popov method in order to handle reparametrization invariance. This work raised much interest and criticism.[i] The main objection was that our gauge depended upon the external sources, and thus is not easily factorizable, in contrast with our previous path integral formulation.[7] We tried hard to understand what was going on, but failed. The answer was given by S. Mandelstam: the light-cone gauge is not conformally invariant, so that there is only one (preferred) parametrization where factorization holds with our gauge fixing. With this parametrization one sees strings (with lengths proportional to their respective p_+'s) which split and join, and our work played a key role for Mandelstam's subsequent discussion of scattering amplitudes, leading to the light-

---

[9]We later published our text in Ref. 11.

[h]This work was initiated during my one-month visit at City College in October 1972.

[i]In particular, from the referee and at a seminar which Sakita gave at the Institute for Advanced Studies in Princeton in December 1972.

cone string field theory.

In the spring of 1973, Sakita visited the Niels Bohr Institute, and I made a trip there to meet with him. On the way back home, he went to CERN and gave a talk at which Zumino was present.[2] Sakita reviewed a current work[13] of Y. Iwazaki and K. Kikkawa[j] who were trying to establish the Nambu-Gotto type formulation of our world-sheet NS-R dynamics. Later on it appeared that this seminar and a subsequent conversation with Zumino played a key role in leading Wess and Zumino to begin their seminal work on supersymmetry. I also remember that the latter author asked me questions about our works on various occasions. After that the whole subject suddenly exploded, and our contribution was temporarily forgotten for lack of reference,[k] given the fact that we had turned to other research directions.

References

1. C. Hsue, B. Sakita, and M. Virasoro, "Formulation of dual theory in terms of functional integrations", *Phys. Rev.* D **12**, 2857 (1970), reprinted in B. Sakita, *A Quest for Symmetry*, Eds. K. Kikkawa, M. Virasoro, and S. Wadia, (World Scientific, Singapore, 1999).
2. See "Reminiscences", in B. Sakita, *A Quest for Symmetry*, Eds. K. Kikkawa, M. Virasoro, and S. Wadia, (World Scientific, Singapore, 1999) and in the present Volume; also in hep-th/0006083.
3. D. Amati, C. Bouchiat, J.-L. Gervais, "On the building of dual diagrams from unitarity", *Lett. Nuovo Cim.* **2**, 399 (1969).
4. J.-L. Gervais and B.W. Lee, "Renormalization of the σ-model. II. Fermion fields and regulators", *Nucl. Phys.* B **12**, 627 (1969).
5. J.-L. Gervais, "Operator expression for the Koba-Nielsen multi-Veneziano formula and gauge identities", *Nucl. Phys.* B **21**, 192 (1970).

---

[j] At that time they were both at City College and had much interaction with Sakita.

[k] Our original paper was nevertheless reprinted in the first volume of Ref. 14.

6. S. Fubini and G. Veneziano, "Duality in operator formalism", *Nuovo Cim.* **97**, 29 (1970).

7. J.-L. Gervais and B. Sakita, "Functional integral approach to dual theory", *Phys. Rev.* D **4**, 2291 (1971), reprinted in B. Sakita, *A Quest for Symmetry*, Ref. 1.

8. J.-L. Gervais and B. Sakita, "Generalization of dual models", *Nucl. Phys.* B **34**, 477 (1971), reprinted in B. Sakita, *A Quest for Symmetry*, Ref. 1.

9. A. A. Belavin, A. M. Polyakov, and A.B. Zamolodchikov, "Infinite conformal symmetry in two-dimensional quantum field theory", *Nucl. Phys.* B **241**, 333 (1984).

10. J.-L. Gervais and B. Sakita, "Field theory interpretation of super gauges in dual models", *Nucl. Phys.* B **34**, 832 (1971), reprinted in B. Sakita, *A Quest for Symmetry*, Ref. 1.

11. *Quantum Field Theory and Quantum Statistics*, Essays in honor of the sixtieth birthday of E.S. Fradkin, (Adam Hilger, 1987), Vol. 2, p. 435.

12. J.-L. Gervais and B. Sakita, "Ghost-free string picture of Veneziano model", *Phys. Rev. Lett.* **30**, 716 (1973), reprinted in B. Sakita, *A Quest for Symmetry*, Ref. 1.

13. Y. Iwasaki and K. Kikkawa, "Quantization of a string of spinning material Hamiltonian and Lagrangian formulation", *Phys. Rev.* D **8**, 440 (1973).

14. *Superstrings, The First Fifteen Years of Superstring Theory*, Ed. J. Schwarz, (World Scientific, Singapore, 1985).

REMINISCENCES [1]

BUNJI SAKITA

Physics Department, City College of New York,
New York, NY 10031

In the spring of 1967 I stayed at the ICTP in Trieste for five months. Towards the end of the stay K. C. Wali and I traveled to Israel, specifically the Weizmann Institute for ten days at the invitation of H. J. Lipkin. When we arrived in Israel we found that the atmosphere was extremely tense and people busy preparing for a war with the neighboring Arabic countries. Although the touristy places were deserted, we could manage to rent a car to visit many places including Jerusalem, Haifa and Acre. Since most of the young Israeli physicists had already been drafted, the physicists working at the Institute were mainly foreigners, among whom were H. Rubinstein, G. Veneziano and M. Virasoro. They were working together on superconvergence relations, which was a subject that I was also interested in at that time. In the discussions we had during this visit, the dual resonance program must have come up, since I remember that afterward in Trieste I started discussing with others about the possibility of constructing scattering amplitudes by summing only the s-channel resonance poles. We left Israel as scheduled on June 4 and the very next day in Ankara, Turkey we heard of the outbreak of the Six Day War.

In 1968, Keiji Kikkawa and Miguel Virasoro, with whom I had become acquainted in the previous trips to Japan and to Israel, joined our group at Wisconsin as research associates. By then I had returned to the University of Wisconsin to resume teaching, which I missed at Argonne, and I was preparing a course, "Advanced Quantum Mechanics", which was essentially a one-year graduate course on quantum field theory. In that summer Virasoro showed up in Madison with a hand-written paper by Veneziano and he explained to us in detail the activities of the Weizmann Institute. At once

---

[1]This is an abbreviated version of the chapter under the same title from B. Sakita, *A Quest for Symmetry* (World Scientific, Singapore, 1999).

Goebel and I got interested in the work and all of us started thinking about generalizations. In that fall after Virasoro succeeded in obtaining the five point Veneziano formula, our activity became intensified and within a few weeks Goebel and I had obtained the N point Veneziano formula. Then Kikkawa, Virasoro and I started to generalize the formula further to include loops.

At this point we faced a dilemma. Namely, if one considered the Veneziano formula as a narrow resonance approximation of the true amplitude as was commonly assumed at that time, the construction of loop amplitudes based on this approximate amplitude did not make sense. After reviewing the logistics of quantum field theory, we arrived at the conclusion that the construction of the loop amplitudes did make sense if we considered the Veneziano amplitude as a Born term of an unknown amplitude for which we had an expansion similar to the standard Feynman-Dyson expansion in perturbative field theory. With this philosophy in mind we decided to construct a new dynamical theory of strong interactions. First we defined the local duality transformation as the crossing transformation at any four point sub-diagram of a Feynman diagram, and invented a Feynman-like diagram compatible with duality as a diagram which contained all the Feynman diagrams related to each other by the local duality transformations. Then we simply wrote down a prescription for the scattering formula corresponding to each of these Feynman-like diagrams. In practice, we used diagrams which were dual to the Feynman diagrams. A three point vertex of a Feynman diagram corresponds to a triangle in the dual diagram. An N point Feynman tree diagram corresponds to a specific triangulation of an N polygon in the dual diagram. A local duality transformation in the dual diagram is the transformation of one triangulation of a quadrangle to another triangulation. In terms of the dual diagram, therefore, an N point Feynman-like tree diagram corresponds to an N polygon. By studying these Feynman-like diagrams, it became clear to us that a dual amplitude corresponded in a one-to-one fashion to a two-dimensional surface with boundaries, and equivalently to a Harari-Rosner quark line diagram, which, by the way, we had also invented independently. In the second paper, we discussed the general Feynman-like diagrams by using the classification of two dimensional surfaces, and extended the prescription to non-planar diagrams. This classification is the same as that of open string amplitudes.

Kikkawa left Madison in the summer of 1969 for Tokyo, and Virasoro and I left the following summer bound for Berkeley and France respectively. By then the operator formalism of the dual resonance amplitude had been established by S. Fubini, D. Gordon, and G. Veneziano, and independently by Y. Nambu, who further proposed the string interpretation based on this work. When I heard of the string interpretation from Nambu I felt it as natural as if I had known about it beforehand. I remember that I had experienced the same feeling when I had first heard about the Sakata model from Sakata.

At Wisconsin Virasoro had used the operator formalism to analyze the possibility of the negative metric ghost states consistently decoupling from the physical states. He obtained a set of operators, which could be used consistently as the operators of subsidiary conditions on the physical states. These operators were later found to be the generators of conformal transformations on a complex plane. This is the origin of the Virasoro algebra. In discussing this problem with him, I realized that these operators were compactly expressed in terms of a scalar field in a fictitious 1+1 space(finite)-time, and the Veneziano formula itself could be expressed in terms of this scalar field operator. At about this time we received a hand-written paper by H.B. Nielsen: "A physical model for the n-point Veneziano model." Inspecting a few mathematical formulae in the paper, I came up with a functional integral representation of the Veneziano formula. There remained several important points to be clarified, such as the Möbius invariant property of the functional integrand, the connection with the operator formalism, and the calculation of non-planar amplitudes. At the end I, together with Virasoro and my student C.S. Hsue, established the functional path-integral formulation of dual resonance amplitudes, and with Virasoro, a physical model of the dual resonance model based on the "fishnet" diagram.

I stayed in France for one year before I moved to the City College of New York in 1971. To lessen the financial burden on the Institute, Michel had arranged a joint invitation by his Institute at Bures-sur-Yvette and the Bouchiat-Meyer group at Orsay, a group of physicists that later moved to the École Normale Supérieure in Paris. This arrangement turned out to be a very fortunate one for me, as in Orsay I found several young physicists, who were interested in our work. Moreover it was there that I succeeded in starting a long and fruitful collaboration with Jean-Loup Gervais. During this visit, I wrote three papers with Gervais: on the functional integral, conformal field theory, and the super-conformal-symmetry, all in connection with the dual resonance model.

In Wisconsin I had already started working on the factorization of dual resonance amplitudes using the slicing and sewing technique of functional integrals. I had drafted the preliminary results into a paper and had sent it to the Physical Review before I arrived in France. At Orsay, however, I withdrew the paper, as a result of discussions with Gervais, when I convinced myself that a part of the paper was wrong. There were plenty of technical difficulties, on which Gervais and I had to spend another half a year of hard work. In this work we used formally and fully the conformal transformation properties of functional integrals without seriously questioning their validity. Sometime later we suspected the existence of an anomaly, that would explain the critical dimension of the model. I regret that we did not pursue it further.

When I received a paper on the new dual pion model of Neveu and Schwarz in the spring of 1971, I noticed at once that the most important ingredient in the model was the conformal invariance property. One could discuss about the generalization of dual resonance amplitudes in the very general terms of conformally invariant field theories. So, Gervais and I got busy constructing conformally invariant field theories. In this work, we discussed first a general theory of conformal fields by defining the irreducible fields (now known as primary fields) and the conformally invariant Lagrangian, and then we established the functional-integral representation of Neveu-Schwarz model by introducing a fermionic field in the model in addition to the old bosonic field. After the work was completed I wrote a letter to Virasoro (in Berkeley then) informing him of our work, since I heard that he had presented a similar work at a conference in Israel. In the exchange of letters, I learned the Ramond model from Virasoro and that it also could be described by the same Lagrangian simply by changing the boundary condition on the fermionic field.

Gervais and I thought that in the functional-integral representation the elimination of ghost states could be done by factoring out the negative metric components of the fields by using conformal transformations as was done in the standard gauge field theories. The necessary condition for this is, of course, that the Lagrangian is invariant under the conformal transformations. Once we introduced a new field in the new model which generated new ghost states, we had to find out a new set of gauge transformations under which the Lagrangian was invariant. Neveu-Schwarz-Thorn had just published a paper in which they proposed a set of operators to be used as the subsidiary gauge conditions on the physical states of the dual pion model. We tried to interpret these operators as the Fourier modes of the Noether current associated with the new gauge transformations which involved the new fermionic field, and arrived at the superconformal gauge transformations, under which the Lagrangian we had obtained previously was invariant. I believe that these field transformations are the first instances of supersymmetry transformations in a local field theory. The day after we had drafted this paper, I left France for New York. In this work, we had to use anti-commuting c-numbers (Grassmann numbers) and functional integration of fermionic variables. These, to us were new concepts and we were initially reluctant to use them. Apparently, others shared this reluctance and this work and the functional-integral work in general, was not appreciated in our circle. However, I received an impression that when I presented the work later in December at the conference on functional integration at the Lebedev Institute in Moscow, it, as well as the use of anti-commuting c-numbers was well appreciated.

When R.E. Marshak became the president of the City College in 1970, I, together with Keiji Kikkawa, accepted a position there. I continued my research on dual resonance theory for a few more years, after I had settled down in the City College. There was a big difference, however, between before and after coming to the City College. Although several faculty members were already there before I came, I was expected to play the role of the leader of the high energy theory group. I felt that it was a great challenge to elevate the group into a quality research group. In a few years, thanks to Marshak's personal connections, we could gather a few talented graduate students into our group. And also we could hire a new faculty member, Michio Kaku and postdocs, such as Yoichi Iwasaki. Moreover, I could invite J.-L. Gervais for short visits on a few occasions. I intentionally spent more time with students, and shared my insights with them.

In the early spring of 1973, I was invited by Ziro Koba to visit the Niels Bohr Institute in Copenhagen for two weeks to deliver a colloquium, and more importantly to discuss the dual resonance string theory with his group, in particular with Holger B. Nielsen and Paul Olesen. By this time at the City College, Gervais and I had already formulated the ghost free Veneziano amplitudes by using the functional-integral representation of the Nambu-Goto string in the light-cone gauge. This work later led to Mandelstam's factorizable functional formulation of light-cone string theory, and eventually to Kaku-Kikkawa's light-cone string field theory. Furthermore in our group at that time, the work of Iwasaki-Kikkawa was near completion. This was an attempt, which I persuaded them to carry out, at a formulation of a light-cone string theory for the Neveu-Schwarz model. I reviewed these activities in Copenhagen. While I was in Copenhagen, David Olive called me up asking me to visit CERN on the way back home. At the CERN seminar, I reviewed the Iwasaki-Kikkawa theory. Later, I was told that this seminar and a conversation after the seminar had led Wess and Zumino to start their seminal work on supersymmetric field theory. I vividly remember the conversation with Zumino at the CERN coffee lounge. When I said, "If you allow me to use anti-commuting c-numbers, Gervais and I have written down a transformation of a fermi field to a bose field in the Nuclear Physics paper", he replied, "It's OK to use anti-commuting c-numbers. Schwinger has frequently used them."

January 1997, New York

2

Discovery of Supersymmetry in Four Dimensions

Y. Golfand

E. Likhtman

D. Volkov

V. Akulov

J. Wess

B. Zumino

NOTES OF AN OLD GRADUATE STUDENT[a]

EVGENY LIKHTMAN

Moscow, Russia

1962: Nine years before the birth of the Superworld in four dimensions. I received a passing grade on my entrance exams to the Department of Physics at Moscow State University but ran into trouble with the medical commission.[b] I gathered my certificates of a winner of many university competitions and arranged an audience with the dean of the School of Physics. I explained to him that I wanted to become a theoretical physicist. His answer was: "No one is above the law".

For me, this was nothing less than a tragedy. Science, and especially physics with all its impressive achievements, was an important and respected field in the Soviet Union. My father "applied" physics with enthusiasm in the defense industry, and we often discussed physics problems together. Requirements for admission to the Institute of Physics and Technology and to the Institute for Engineering and Physics were even tougher than those for the physics department at Moscow University.

My teachers arranged for me a meeting with the President of the University. During our interview, I told him about an inaccuracy I had noted in one of the problems in the entrance exam. As a result, once I had passed the entrance exams to the so-called extension division of the Department of Mechanics and Mathematics, I was by order of the President, transferred after one month to the Department of Physics.

In the course of five and a half years of study there, I learned how to take the traces of the products of gamma matrices, learned something about Clebsch-Gordon coefficients, and was fascinated by the beauty of the Noether theorem. I completed my senior project under Yu.M. Shirokov, who worked at the Mathematics Institute of

---

[a]This essay was first published in the Yuri Golfand Memorial Volume *The Many Faces of the Superworld* (World Scientific, Singapore, 2000).

[b]Here and below see Editor's comments collected at the end.

the Academy of Sciences. However, the Graduate School of this Institute turned me down, and the advisor of my senior project sold me to the Theory Department of the Lebedev Physics Institute (FIAN). This was kind of an "arranged marriage". Thus, it was that at the right time, in the spring of 1968, I found myself in the right place: working with Yuri Abramovich Golfand.

Yuri Abramovich (for that is how I always addressed him, the difference in our ages being more than 20 years[2]) was somewhat short, with long arms that he waved around as he walked, rather like a skier or a speed skater. He was always in a good mood and never raised his voice. And whenever, as it happened, I made some blunder, his face would become radiant — this was a gradual process — with an enchanting smile. I often observed how other people were completely disarmed by this smile of his.

Twice a week, one hour before the beginning of the seminar, I came to the small room of the Theory department with its four desks, and we discussed my results. I wrote out my calculations on the back of the blueprints of old drawings, which were rolled up like wall paper. Bystanders observed with interest the process of rewinding these rolls. Golfand did not always stay for the seminar after our conversations, he would sometimes slip away with a young female graduate student. I, on the other hand, was very concerned about the development of superscience and strongly disapproved of his lack of dedication.

Superalgebra had been written by Golfand before my appearance in the department. I had to figure out whether less complex superalgebras existed and then to determine whether they had any relation to field theory or high energy physics. The first part didn't take much time — I wrote out fairly quickly all extensions of the algebra of generators of the Poincaré group by bispinor generators. It took significantly longer to put together free field representations: one had to get used to the fact that in one multiplet were unified fields with both integer and half-integer spins. I built a series of representations which included high spins as well. This was a result of some kind, and I wanted to get it published as soon as possible. But instead, Golfand set me the task of building theories of interacting superfields.

1970: The term "supersymmetry" did not yet exist. We didn't know anything about supergroups or superfields with Grassmann parameters. Superinteraction was constructed by the method of indefinite coefficients: the most general form of interaction was written (up to the quartic terms in the fields) and spinor generators (to the third power of fields), after which these polynomials were substituted in the superalgebra, and then the equations determining numerous unknown constants of polynomials were written down with the goal of finding solutions. Among the equations, there were many dependent ones, and this gave hope.

By this time Golfand had finished his doctoral dissertation[3] and ...went off on a ski trip. By doing this, as it seemed to me then, he showed a lack of responsibility and also a certain disregard for appearances.

The first path to the superworld was not the shortest. Now I can no longer remember what specific directions Golfand gave me. More important was the stable, symmetric foundation of the theory and his apparent unshakable faith in its correctness. Superinvariant interaction was eventually constructed,[4] and the time came to publish the results. I remember how we got hold of A.D. Sakharov in the corridor of the Theory Department and asked him to recommend our work to the *Reports of the Academy of Sciences.* He talked us out of this idea because of the time-consuming review process in this journal. The *JETP Letters* accepted only short articles, and I wanted badly to set everything out in detail. Golfand abridged my manuscript with a stern hand. Free field theory came out[5] as *Preprint FIAN #41.*

Two reviewers of my candidate dissertation — V.G. Kadyshevskii and V. Man'ko — made efforts to understand my computations in detail. The results of our work were presented at FIAN, the Institute of Theoretical and Experimental Physics, Moscow University and the Joint Institute for Nuclear Research. A detailed article was also written for the I.E. Tamm Memorial Volume.[6] I defended my dissertation without much ado. At the banquet, Golfand arrived with a big bouquet of flowers, which, to my surprise, was meant not for me, the trailblazer and hero, but for my wife, who stayed at home with our newborn son.

After graduate school, I was rejected by the Theory Department of FIAN, as well as others. The only place where I could get a job was the All-Union Institute of Scientific and Technical Information (VINITI). A certain theoretician, who heard about my place of work, made an analogy to Einstein. At my Institute, conditions were ideal for scholarly work: official rules did not prevent me from doing research once a week.

For Western readers it should be noted that in the Soviet Union one took a job more or less for life. To fire someone legally, three warnings and the agreement of the union were necessary, or ... elimination of the position. In 1972 the Academy of Sciences was required to make personnel cuts. It was a routine campaign. As a rule, administrators of scientific institutes eliminated only unfilled positions, retired those who had reached pension age, and found various means of retaining actively working colleagues. In the Theory Department of my Institute, Golfand was designated expendable. As Sakharov wrote later[7] in the popular magazine *Novii mir*, the dismissal of Golfand was unique at FIAN. I argued with administrators for Golfand's retention but was rebuffed.

1973: Now our less frequent meetings occurred first at one, then at another of Golfand's apartments on Lenin Prospect, not far from the Academy of Sciences. Golfand remarried; his young wife looked at him adoringly. After trauma from a skiing accident, he developed a slight limp but didn't like to talk about this. He professed good cheer, and I used to tell him all kinds of things I learned from the grapevine. Only his occasional comments betrayed his attitude toward his former colleagues at the Theory Department of FIAN.

I continued, with the help of obtained techniques, to construct new versions of interactions of the supermultiplets. But all these were far from the real world of elementary particles. I was the first to observe that the number of the fermion and boson degrees of freedom coincided in every supermultiplet, while the infinite energy of vacuum oscillations is cancelled. This discovery did not make much of an impression on those around me. Later I proved that one-loop boson mass diverged not quadratically, but, like the fermion mass, only logarithmically. Even more importantly, I succeeded in constructing a renormalizable theory of massive vector fields. This was not "good news" in the scientific community, since in the Weinberg–Salam model this problem had already been solved. Of importance were nonrenormalizable theories which supersymmetry would make renormalizable. But for the solution of this problem one needed a different technique for constructing superinteraction.

1974: Supersymmetry has been rediscovered by Western scholars. Superfield representation easily allowed them to construct superinvariant interactions. At my institute, VINITI, I was collecting copies of works on supersymmetry, but my file quickly overflowed. I came to the conclusion that I had to find my own topic.

1985: Peter West proposed we write an article for the collection *Supersymmetry: A Decade of Development*. Golfand had something to say about this subject, and he become absorbed by the idea and dictated to me his conditions: he would write the text of the article in English, and I would translate it into Russian for GLAVLIT.[8] During this work he found it very difficult to fulfill all the standard requirements for publication. I remember with what apprehensive cautiousness, fearing refusal, Golfand asked the publisher to alter requirements for the preparation of the manuscript.

1989: L.V. Keldysh — who was head not only of the Theory Department but of the entire Lebedev Physics Institute — nominated our work on supersymmetry for the Academy of Science's Tamm Prize in theoretical physics. The presentation of the award occurred in the impressive setting of the assembly hall of Moscow University. Golfand, accompanied by his wife, and I arrived separately and sat in different parts of the hall. The voice of the Academic Secretary snatched up people to the left and right of me, and they went up to the stage to receive their awards. Then came our turn. I was naturally the second to step forward, and Golfand did not see me. After he received the award, he asked the President of the Academy if he had seen Likhtman. I was struck, yet again, by his childish ingenuousness.

1990: Golfand obtains documents for immigration to Israel with his wife and two lovely daughters. Our farewell dinner in the kitchen.

And, as a memento, an album of formulas I had studied back in 1968.

And then I turned away from the mainstream of high energy physics, where symmetry conducts a great orchestra of fields, and turned onto a different path with a small ensemble of prefields that form extended soliton-like solutions. About the same time I staked my hopes on personal computers, that began appearing at VINITI in the late eighties, and on my developing skills in numerical solutions of variational problems. Within the framework of the Born–Infeld model of nonlinear electrodynamics, I hope to determine the dimensionless characteristic of a soliton that would be independent of the parameters of the model, which is trivially connected with the fine structure constant. But that is another story . . .

1991: Golfand returns to Moscow for the Sakharov conference. At our meeting, he could not conceal his disappointment with life in Israel, which began with a long search for lost baggage at the airport. He was forced to deal with indifferent clerks, their "European gloss" notwithstanding, as he put it. Despite the fact that the Minister of Science of Israel certainly knew about his contribution to high energy physics, Golfand was unemployed for a long time and lived on welfare. At the time of his trip to Moscow, his life in Israel was just beginning to take shape. And once again our conversations were dominated by his beloved theme: the prospects for receiving a Nobel prize for the discovery of supersymmetry. I was again shocked by his superoptimistic expectations. Now, in hindsight, I understand that despite his seeming autonomy, and scorn for appearances my Teacher desperately needed the support and recognition of other physicists. On the other hand, here again was apparent his unrestricted range of thought in posing superproblems, and this was, perhaps, the last and most important lesson I learned during our work together.

The original written in Russian by Evgeny Pinkhasovich Likhtman in March 1999, was translated in English by A. Liberman and M. Shifman.

Commentary (composed by M. Shifman)

1. This was a standard trick allowing the Soviet authorities to keep quota on Jewish students in universities without ever admitting that the quota existed. The first barrier was the entrance exam itself. Undesirable persons were offered problems which were much harder than those given to everybody else. If, against expectations, a given undesirable person was successful in solving them, the medical commission would always find a pretext to declare him/her unfit for studies in the given university for medical reasons.

2. Formal address in Russian is by name and patronymic, rather than by surname.

3. The academic hierarchy in Russia follows the German rather than the Anglo-American pattern. An approximate equivalent of PhD in the US is the so called *candidate* degree. The highest academic degree, doctoral, is analogous to the German *Habilitation*. The doctoral dissertation is usually presented at a mature stage of the academic career; only a fraction of the *candidate* degree holders make it to the doctoral level.

4. Yu.A. Golfand and E.P. Likhtman, *JETP Lett.* **13** (1971) 323 [Reprinted in *Supersymmetry*, Ed. S. Ferrara, (North-Holland/World Scientific, Amsterdam – Singapore, 1987), Vol. 1, page 7].

5. A. Likhtman, Report of the Lebedev Physics Institute # 41, 1971, in Russian. On page 8 of this preprint one can read, in particular: "As is known, in relativistic quantum field theory, in transforming the free energy operator to the normal-ordered form there emerges an infinite term which is interpreted as the vacuum energy. It is also known that the sign of this term is different for the particles subject to the Bose and Fermi statistics. The number of the boson states is always equal to the number of the fermion states. From this it follows that the infinite positive energy of the boson states in any of the representations of the [super-Poincaré] algebra is annihilated by the infinite negative energy of the fermion states."

6. The English translation of this paper, *On the Extension of the*

Algebra of Generators of the Poincaré Group by Bispinor Generators, I. E. Tamm Memorial Volume *Problems of Theoretical Physics*, (Nauka, Moscow 1972), page 37, is published in *The Many Faces of the Superworld*, Ed. M. Shifman (World Scientific, Singapore, 2000), p. 45.

7. Here E. Likhtman refers to the memoirs of Academician Andrei Dmitrievich Sakharov. English Edition: A. Sakharov *Memoirs* (Knopf, New York, 1990), translated from Russian by Richard Lourie.

8. GLAVLIT was an agency implementing total censorship in the USSR. Every material intended for publication, including such insignificant stuff as, say, the business cards and the like, had first to be cleared through GLAVLIT. The GLAVLIT bureaucrats would require from the authors to supply them with many copies of the paper, in Russian, even in those cases when the paper intended for publication in one of the Western journals had been originally written in English. Since there was no access to photocopying machines, this was a hard task by itself. In general, the process was time and labor-consuming, and I heard from Golfand on several occasions that he hated it.

MY HUSBAND, MY LOVE[a]

NATASHA KORETZ-GOLFAND
Maale Adumim, Israel

Writing this was not easy, for despite the fact that Yuri passed away over five years ago, I still love him. Emotions overwhelm me, making it hard to write something logical. That's why this turned out so chaotic and fragmentary. But I am sending it anyway...

I first met Yuri at the end of 1961, when I was almost 16, and he was almost 40. He was married, and his oldest son was only one and a half years younger than I. Despite all that, I fell in love with him from the first sight, and tried, as much as I could, to excite his interest, gain his attention, and to be close to him even just as a friend. Intuitively I realized right away that we were an ideal couple. For me, it seemed impossible not to see that we were meant for each other, that we had to be together in grief and in happiness and for the rest of our life...

Later on, that is just how it came to be. But for seven years (as in the Bible) we remained friends and nothing more. We belonged to the same "hiking posse." We went to the mountains, canoeing, skiing. But in everyday life we didn't see each other very much.

Yuri worked a lot; he was an excellent theoretician. There were legends about him that he could "crack" any problem in three days, well... perhaps in four, if it was particularly difficult.

A first-class physicist, athletic, very handsome, not only did he like to go skiing, canoeing, and hiking, but he was also one of the first people in Moscow to start practicing yoga. He procured translations and *samizdat*[b1] editions of guidebooks, very rare at the time, and taught himself hatha and raja yoga. He knew a lot about Soviet and foreign literature, poetry and art. He was a big fan of *bards*.[2] Bulat Okudzhava, whom he called "the Teacher," and Alexander Galich. I remember how he copied their songs onto his tape-player and later,

---

[a]This recollection was first published in the Yuri Golfand Memorial Volume *The Many Faces of the Superworld* (World Scientific, Singapore, 2000).

[b]See the comments collected at the end.

to the accompaniment of a guitar, sang them with us by the bonfire.

He was contagiously young, younger than all of my peers, and at the same time it was always interesting to be with him – he always reflected on deep and unusual matters.

Where he found time for all this is difficult to comprehend. To many of his friends and acquaintances he seemed to be a "Mozart in science." His life appeared to be pure enjoyment; but then, "suddenly" and with no apparent efforts, he would publish a new scientific paper. This paper would undoubtedly contain an unexpected and original physics idea which would be later picked up and worked on by many scholars in his and neighboring fields. That was the case with his theory of supersymmetry, – so new that no one in the theoretical department of FIAN[3] was able to understand it for quite some time. Both in science and in life, Yuri was seriously interested only in matters of substance – new and not yet understood by the others.

Yuri had so little tolerance for Soviet propaganda that it could make him physically ill. The "garbage" that came from the media, the "noise," as he called it, wore him out immediately. Because of that, in life, in art, and in his relationships with people, he was very demanding.

But what surprised and attracted me the most was not Yuri's brilliance and erudition, but his absolute freedom in everything. He was born and raised in a totalitarian country; he was educated in regular Soviet schools and universities. He came from a wonderful family and had a relatively happy childhood, at least in contrast with my own. (I was born in one of Stalin's *Gulag* camps, and till the age six lived near "the zone."[4] My parents had been arrested in the infamous 1938, and were able to come back to Moscow only 20 years later, after the 20th Party Congress[5]) Despite his happy childhood, however, Yuri found political hypocrisy and any other lies just as intolerable as I did. He never lived "according to the circumstances," as the majority of the Soviet intelligentsia was forced to live, finding an escape in the "theory of small deeds." According to this theory, one could be a decent person and at the same time remain a loyal Soviet employee (which was the only way to do the work they liked). Many tried to escape into "pure science," "pure art," and narrow professionalism.

Yuri was not afraid *not* to join the Communist Party. He refused to participate in the yearly campaigns of "assistance" to the kolkhoz farmers, when all of Soviet intelligentsia – from university students to renowned professors – were sent to the fields to pick carrots and potatoes.[6] He was not afraid to read and keep forbidden literature (books by Solzhenitsin, Daniel, Brodsky), which was a criminal act according to the Soviet law. He did only that which he wanted to do. At the time many people considered this a defiant and unaffordable luxury, but Yuri simply couldn't live otherwise.

When his friends became dissidents or even were arrested, Yuri didn't break off contact with them. On the contrary, he offered help to them and to their families. After the arrest of the physicist and human rights activist Yuri Orlov (Golfand's old friend and colleague), Yuri participated in the protest pickets against the trial, where no one but the closest relatives and the specially selected KGB "guys" were allowed to attend. He later visited Orlov in exile, accompanying his wife Irina Orlov from Moscow to Kobyai in Siberia and back. Yuri wrote an article about Orlov. He did all this not only to expose the injustice committed by the authorities towards a gifted scholar, but also to promote Orlov's elegant idea of wave logic as it so deserved.

In Yuri's system of values, the notions of decency, honor, and integrity were an indispensable part of existence, despite the fact that these values were being systematically eradicated from the public consciousness by the Soviet regime. Yuri used to say that decency was like fish: it could not be "half-fresh." But in Soviet stores fish was sold not only half, but a "third-fresh," and even for that, one had to stand in a long line. Yuri would never buy such fish, and if there was nothing else, he could go on bread, refusing to adjust to inhuman conditions.

At the time when we became *refuseniks*,[7] there was, among the other Moscow dissidents and *refuseniks*, a popular handbook written by Volodya Albrekht[8] on how one should conduct oneself during a KGB interrogation: how to answer or not to answer their questions in order to avoid bringing harm to oneself or to anybody else. This handbook was very popular, for the KGB used these "conversations" to intimidate dissidents. Although the interrogations did not follow any legal procedures and had no judicial grounds whatsoever, to decline the invitation was extremely dangerous and could become the reason for one's arrest. Yuri was invited to these "talks" several times. Not hesitating for a single moment and not afraid of the possible consequences, he would refuse to come to the "meeting" and hung up right away.

We were first denied the right to emigrate to Israel in October 1974, after having submitted the application in December of '73. For almost a whole year we lived in uncertainty, which was unpleasant, to say the least.

Yuri began to learn Hebrew. He brought home a textbook and a tape and studied every day for an hour. At that time I was finishing my studies at the university and was extremely busy. But I also tried not to fall behind. In two months, Yuri could read and write, knew the rules of grammar and tried to listen to the news on the radio, while I was still forming simple sentences!

He was just as fast at mastering computers when he later started working at Haifa's Technion. In two weeks, he learned to work on a computer and used it to lead his correspondence and type all his papers.

There is a stereotype in the public eye of a great scientist fit only for scientific matters and utterly helpless in all practical affairs. Yuri could do almost everything by himself, with his own hands – from cleaning the house to cooking to serious repairs of electrical systems and appliances. He bragged that in him was "dying" a famous plumber. When we went canoeing or hiking in the mountains, I felt myself as if behind a stone wall. There was never a problem to start a fire in the rain or to patch up a torn coating of the canoe – Yuri always had superglue and dry fuel handy.

When our two daughters were still babies and required constant attention, Yuri helped me in everything – he washed them, changed diapers, did laundry. He took pride in being able to replace me even in nursing them (there was no baby food in Moscow at the time, so that if I was going out, I had to leave him a bottle with my own milk). But when he sat down to work, he would not be distracted by anything. He organized his space not only on his desk, but in the whole house. He said that the work of a scientist is no easier than that of a ditch-digger – before reaching an elegant result, one had to turn and haul over a huge pile of rocks. Indeed, after finishing an article of 10 –12 pages, he would throw away 200 – 300 (sometimes more) sheets of rough drafts.

Yuri knew his worth as a physicist, adequately appraising his contribution to the quantum field theory, realizing that his theory of supersymmetry would forever remain a classic in the history of physics. But he never showed off his achievements and disliked all self-advertisements.

When we finally immigrated to Israel, Yuri was over 65. Many of his friends warned him that he wouldn't be able to find a job: in 1990 the rate of *aliyah* to Israel was at its peak. We heard Mark Azbel explain on the radio that the newly arrived and still unemployed scientists should thank Israel for letting them sweep the streets.

Yuri joined *Ulpan* at the most advanced level and actively began looking for a job. He sent out a list of his publications, gave talks at the universities of Tel-Aviv and Jerusalem, at the Weizmann Institute, in Technion. He got help from his *refusenik* friends, from scholars who had immigrated earlier – Victor and Irina Brailovsky, Alexander Ioffe, Alexander Voronel. During these difficult times, we got financial support from the Israel Public Council for Soviet Jewry. Six months after our arrival, with the help of Yuval Ne'eman and Misha Marinov (both of whom understood Yuri's significance as a physicist), Yuri received a position as a research fellow in the Physics Department of Technion. There he composed drafts of many new physics papers, but had time to complete and publish only a few of them. He just began **the** work, that he considered most important for himself; he even found a mathematician – Dov Ramm, with whom he was going to develop a mathematical apparatus.

I am a Humanities person, and the physics of elementary particles is beyond my reach. Nevertheless, Yuri was somehow able to explain to me the subjects of his work. Despite the difference in our occupations, we always found it very interesting to be together. We discussed all kinds of things. During our first years of marriage we would even get into arguments about books or movies, until we finally agreed that everyone had a right to their own opinion.

Yuri thought a lot about the structure of the world – he wondered who created it and how. In a way, all his work was an effort to break through to that kind of knowledge.

Through his practice of yoga, he mastered incredible things – transcendental journeys, controlling his pulse and his blood pressure (he would use this ability during medical exams, when he had to get medically cleared for a skiing trip in Chiget).

Once, still in Moscow, he tried to tell me that the visible reality was only a reflection of our multidimensional world, as a shadow of a lamp post on the ground. We perceive only the shadow and have no idea about the post or the light. (This happened during one of our evening winter walks. On the whiteness of the fresh snow, the shadow from the lamp post appeared very defined and glowed under the bright light). At the time I couldn't comprehend or accept his idea. It frightened me and for some reason seemed inhumane.

After Yuri's death I had the opportunity to read several serious books on mysticism. I was shocked to find great similarity between the model of the universe and of humanity described in the books and Yuri's own ideas. There is no possible way that he could've found such literature in Moscow, so he had to have discovered them completely on his own. Back then, I got scared and closed myself off from this kind of information, but now I deeply regret that I have failed to understand him right away.

Now, when I try as much as I can to comprehend these matters, I become more and more aware of the uniqueness and the power of my husband's personality, and I thank God for the happy and the difficult 32 years that I could spend close to him.

Our life was indeed very happy and very difficult. We had enough trials and tribulations that fell on our path. The *otkaz*[7] years were hard not only because we had to stand against the regime, not only because there was no work and no money. For Yuri, the hardest thing was the isolation from his colleagues – the physicists of FIAN. The people with whom he had worked together for 20 years cut off all communication with him, afraid of approaching a dissident ostracized from the official "Temple of Science."

I don't like to recall all this. Nearly 25 years have elapsed since then. But at the time, the indifference and the cowardice of his former colleagues deeply wounded Yuri. On top of this, at approximately the same time, Yuri had to break off contact with his children from his first marriage, who could not forgive their father for leaving the family. (Thank goodness, this break turned out to be only temporary.)

All this had shaken Yuri up immensely, so that he couldn't work for almost three months. What eventually got him through was his yoga practice and the people who have stuck by us despite everything. We were surrounded by beautiful people, thanks to whom we didn't become completely isolated. Yuri's mother and sister, their friends, my parents, their friends and mine – those people were few, but they were not frightened by our new "status" and tried to support us in every way they could.

In some time, Yuri got introduced to other *refusenik* scientists, who united together to hold seminars. That's how our 16-year-long life in *otkaz* began: talks and conferences with scientists from all over the world; rallies of protest demanding the authorities to let Jews out of Russia; communal celebration of Jewish holidays; house search and arrest; participation in the publication of the magazine "Jews in the USSR;"[9] arguments about the immigration of Jews to Europe and the United States...

Although our Jewish life was very eventful, Yuri's work in physics remained his top priority. He also found happiness in our two growing daughters. He continued to go hiking, although now less frequently and only to the outskirts of Moscow – there was no money for longer trips. He still worked a lot. Thank God a theoretician only needs his head, a desk and some paper to do his job. Of course, he lacked contact with other physicists, but even without it he was able to write and publish several scientific works. During these years Yuri became a member of the New York Academy of Sciences, of the European and American Physical Societies. The most surprising recognition, however, came from the Soviet Academy of Sciences in 1989, when Yuri and his former student E. Likhtman were awarded the I.E. Tamm Prize for their works on supersymmetry.

Our life in *otkaz* not only made us stronger, but it also cured us from nostalgia for a long time afterwards. Orwell's *1984* gives only a bleak idea of what we went through in dealing with OVIR.[10] I had to subject myself to the torture of interacting with Soviet bureaucrats; Yuri could not stand their arrogance, and refused to reconcile himself to it.

For many years I demanded to know why we were denied the exit visas. The answers that I got were all the same:

– This is not in the interests of the Soviet Union, – was the most frequent answer.

– Divorce your husband and get the hell out of here, if you'd like. But as long as your husband is alive, he is not going anywhere, – was the rudest and the most sincere answer.

I would like to forget. Forget forever the humiliation of endless lines in the halls of OVIR. Forget the time when they took away our passports, making us pay a huge sum for *renouncing* the Soviet citizenship. Forget how we had to spend several sleepless nights at the airport, trying to check in our luggage; and how nonetheless, with the exception of skis and linens, none of our luggage ever made it to Israel... All that is over now. It's been almost ten years – it's time to forget it once and for all... Only Yuri's mother and my parents didn't make it to the day we were finally allowed to leave. They passed away when we were still *refuseniks*. That, we cannot forget. That remains our pain forever...

The difficult years in Russia were followed by the problems of adjustment to our new life in Israel. Of course, these were problems of a different kind, but it was still hard, especially for Yuri. He could not get accustomed to the heat and the humid winds; could not get used to the different landscape, to the small distances between the cities, to the absence of rivers which he so dearly loved back in Russia.

He did not want to adjust to Israeli bureaucracy, and to the "small-town" mentality of Israeli people. But contrary to all expectations, the ultra-orthodox Jews in their theatrical costumes did not bother him. Perhaps it was in those people, and not in the secular part of the population, that lay the reason for the existence of a country like Israel, Yuri thought. Naturally, the fanaticism of the orthodox was alien to him, just like any other form of fanaticism. But he read the Torah and dreamt to read the teachings of the Kabbalah. One day, he hoped to find a very intelligent and not fanatical religious person, who would agree to discuss certain **untrivial ques-tions** about the religious notions of the world's existence. – That, he didn't find...

He worked, as always, very hard. During the last six months he began to work on a new idea which he imagined to be even more interesting and powerful than the idea of supersymmetry. In the midst of realizing his plans, he had a stroke. Three days later – on February 17, 1994 – he died.

It was impossible to believe that he was no longer with us. Everyone was in shock: his friends, relatives, colleagues. Many people came to say good-bye; those who couldn't make it sent their condolences.

For a long time I believed that it hadn't happened. At night in my dreams, I saw the continuation of our life together. But recently I started feeling that Yuri's beautiful soul is in another world, whatever it could be, and is involved in something very important. He seems fine there, and I am at peace. But I still miss him so...

The original was written in Russian by Natalya Moiseyevna Ko-retz-Golfand in June 1999, and translated into English by Anya Shif-man.

Commentary (composed by M. Marinov, A. Shifman, and M. Shifman)

1. *Samizdat* – A huge variety of books were forbidden for publication (or restricted in publication) by Censorship. The class of disfavored books included not only those with political connotations, but, in general, all books which were considered not helpful for the Soviet ideology. Copies of forbidden books were being smuggled in the country illegally. People retyped them, using mechanical type-writers and carbon paper, or photograph them, page by page, using amateur cameras, and then print at home on photopaper, producing huge piles. The process was called samizdat, which can be loosely translated from Russian as self-publishing. *Samizdat* was forbidden by the Soviet law.

2. *Bards* – This word exists in English too, meaning a poet and a singer. In the Soviet Union it was an important cultural phenomenon, beginning in the 1960's. The bards composed their songs, both music and lyrics, presenting a sharp contrast to the everyday lies of the official propaganda. They sang their songs to small groups of people at private apartments, usually accompanied by a guitar. The songs were taped and then re-taped many times, from one tape recorder to the next. The best of the bards – Bulat Okudzhava, Alexander Galich, Vladimir Vysotsky and others, were the pride and honor of the nation. People found refuge in their songs – about life and love, honesty and integrity – from the absurdity of the existence in the communist regime. Distributing such tapes was formally outlawed as anti-Soviet propaganda. Alexander Galich was forced into exile, where he shortly died under suspicious circumstances.

3. FIAN – Lebedev Physics Institute of the USSR Academy of Sciences, in Moscow.

4. *The Zone* – In popular jargon, the inner part of the GULAG camps where the inmates were kept in barracks.

5. XXth Congress (of the Communist Party of the Soviet Union) took place in 1956. At this Congress, at a closed meeting, Khrushchev gave a speech revealing and denouncing the mass terror of Stalin's rule. The phenomenon was ascribed not to

the general nature of communism, however, but to the so-called Stalin's personality cult. After the XXth Congress many of the GULAG camps were shut down, and millions of innocent people released.

6. *Potato picking* – Each fall all those who were considered unproductive – college students, engineers and scientists, musicians and so on – were sent for a month or so to "help" disintegrating Soviet kolkhozes (collective farms) to collect crops and save at least some of them for the winter. In popular jargon, this routine was commonly referred to as "potato picking," although in practice this could mean any type of agricultural work. As with any slave labor, it was extremely inefficient. The villages usually could not accommodate crowds of people sent there from the cities – the conditions of existence of these *ad hoc* farmers were substandard. This metayage was mandatory; refusal to go "potato picking" could cost one his/her job or place at the university.

7. *Otkaz* (from the Russian "refusal") – the name for social and political conditions under which a large group of people (the so-called *refuseniks*) were forced to exist in the USSR in the 1970's and 80's. The only "crime" committed by these people was that they had applied for and got denied the exit visas to Israel. And yet, they were treated essentially as criminals: fired from jobs and blacklisted, with no access to work (with the exception of low-paid manual labor), constantly intimidated by the KGB, at the verge of arrest. In fact, the most active of them, those who tried to organize and fight back for their rights, were imprisoned.

8. V. Albrekht, a dissident and human rights activist, wrote an article explaining how dissidents could keep their human dignity and resist pressure of the KGB, without entering into conflict with the Soviet laws. The article was distributed as *samizdat*.

9. *Jews in the USSR* – collections of articles on Jewish history and culture, and on important current events. In a sense, this was the chronicle of the Jewish community: information on the struggle of Jews for their rights, on arrests and other persecutions, on *refusenik* seminars and so on, was provided on a

regular basis. Over several years, these collections were issued and distributed as *samizdat*. This was one of the successful endeavors of *refusenik* community. A reprint of the full set is available now (the Library of the Jerusalem University).

10. OVIR – Department of Visas and Permits, a section of the Ministry of Interior Affairs which was in charge of exit visas.

LAST INTERVIEW WITH D.V. VOLKOV[a]

D.V. VOLKOV

Theoretical Laboratory, Kharkov Institute of Physics and Technology, UA
310108 Kharkov, Ukraine

Q. Dmitry Vasilyevich, could you tell me, please, what was your path into theoretical physics, how did you become a theoretician, and was this accidental or was there any specific cause.

A. I was born in Leningrad. When I was 16 years old and when I was studying in the 8th grade, the Great Patriotic War began. I was evacuated from Leningrad. These were very difficult years for young people. In this period I came to work on a collective farm and in a military factory. After that I was drafted into the army and took part as a soldier in the war on the Karelian front, above the polar circle. When the war began with Japan, I participated in military action on the Far Eastern front. I want to say that the war had a considerable influence on my attitude to life. In my generation the war created a feeling of responsibility for the country. After the war we carried over the same ideology into civilian life. When the question of a choice of profession arose, many of us thought about how we might be useful to our country. During all the war years I dreamed about going into science, because already in school I was attracted especially to the exact sciences: mathematics and physics. After the demobilization I entered Leningrad State University, in the faculty of physics. At that time prominent scientists such as V.A. Fock and T.P. Kravets were teaching there. The lectures of T.P. Kravets were distinguished by his ability to link the study material with personal moments. He taught us that physics is created by living people and he spoke much about his teacher Lebedev. From the first days Kravets infected us with a deep love for science, and for physics in particular. Aside from that, I listened to the lectures of V.I. Smirnov, whose widely known multi-volume works were specially intended for theoretical physicists, and, in fact, formed the basis of our whole education. We learned

---

[a]Questions by Yu.N. Ranyuk, reprinted from *Supersymmetry and Quantum Field Theory*, Eds. J. Wess and V. Akulov (Springer Verlag, Berlin, 1998), pp. 1 – 8, with abbreviations. Translated from Russian by Prof. K. Stelle.

a lot from other mathematicians of his school: O.A. Ladyzhenskaya, M.I. Petrashen. In the final year I received a profound training in the specialization of theoretical physics thanks to the excellent teacher L.E. Gurevich. There were also other teachers, which I remember to this time with thankfulness.

Unfortunately, a great tragedy happened in Leningrad when I was studying there, and this had an impact on my further destiny. At the end of the 1940's the so-called Leningrad trial took place, as a result of which prominent [Communist] party leaders and representatives of the scientific and cultural community of the city were repressed, and among them were the leading scientists of the Leningrad University. Therefore, the group in which I was studying was dissolved and the scientific line on which I was working ceased to exist there. Here one should give credit to the Kharkov scientists. From the very beginning, the Kharkov scientists had close contacts with their colleagues in Leningrad. In particular, it is known that our Institute, KhFTI, was organized thanks to the arrival in Kharkov of the group of scientists headed by the well- known physicists Kirill Dmitrievich Sinelnikov, Anton Karlovich Walter and others. One of the major concerns at the foundation of the Institute was how to attract young people so that in the future the Institute would be staffed by highly qualified scientists. Therefore a special representative came to Leningrad, to recruit students who were offered the possibility to move to Kharkov and to continue their studies there. I agreed. Together with me came the already well-known scientists E.V. Inopin, K.N. Stepanov, V.F. Alexin, and some other students.

Q. Dmitry Vasilyevich, when did you move to Kharkov, and how did your further scientific interests develop?

A. I moved to Kharkov in 1951. Now, about my scientific interests and their development. During my student years, the mathematical aspects of science attracted me even more than the physical ones. At that time I had already developed a deep feeling for the general theory of relativity. Quantum field theory was just beginning. Just at this same time the development of physics received some sharp impacts. On the one hand, quantum electrodynamics was effectively reborn thanks to the theory of renormalization, and, on the other hand, the physics of new particles arose. At this time the π meson was observed and the neutral π^0 meson was discovered and discussions on the definition of spin and parity were going on. This was a real revolution in science. Later on, after graduation from university, I began my graduate training. My advisor was A.I. Akhiezer. He organized a small group of graduate students, including R. Polovin, P. Fomin, V. Alexin and me. All of us actively studied quantum electrodynamics. This was especially important for me, because I continued to work in this area, and quantum electrodynamics became for me a sort of initial example of the theories that I work on now.

I would like to discuss in some more detail the directions of science that interested me most and on which I worked later. I don't know why — somehow intuitively — but already in my years of study my favorite subject was connected with the theory of symmetry groups. At that time, the theory of symmetry groups was already being adopted in physics, but not widely enough. Later on, the application of this theory came into a full flowering, especially when many new elementary particles were discovered and the question of systematizing them on the basis of symmetry groups became very important. When I started to work, there were no powerful group-theoretical methods. The first pieces of work that I consider to be somehow connected to what I am doing now were concerned with the properties of particles with high spin. I would like to emphasize that questions that were also related to supersymmetry interested me already at the earliest stage of my activity. How do particles with different spin differ from one another? Why are there bosons, why are there fermions? I remember, in part, that just when I learned about the group SU(3), it amazed me that the multiplets of this group contained simultaneously particles with integral and with half-integral isospin. I was already trying to do something in this direction, replacing isotopic spin with ordinary spin, that is, actually, the idea that now lies at the basis of supersymmetry. An important role in the conception of the idea of supersymmetry was played by work on phenomenological Lagrangians, in which the interactions of Goldstone particles in the presence of spontaneous symmetry breaking are practically uniquely described by geometrical group-theoretical methods. I actively participated in the development of this line; it fascinated me very much. At the same time, the ideology of gauge field started to penetrate actively into physics and I returned to the old questions. It amazed me that all these particles, the Goldstone particles and the gauge fields, were bosons, but the fermions somehow were not involved at all. Here, a kind of inequality appeared: why were some particles — bosons — selected, but others — fermions — not included in this group? This was a key moment, because the very thought that fermions can also be Goldstone particles or gauge fields contained in itself somehow the answer: if it were clear how to build a general scheme of Goldstone particles with integral spin, then upon making the transition to fermions one should replace the corresponding operators with operators firstly carrying spin one-half and secondly being anticommuting, in accordance with their fermionic nature; and in fact I did this. After that, when this was done, I, together with my co-authors V. Akulov and V. Soroka, considered first the global properties of supersymmetric theories and then local properties.[1,2]

I would like to say that I worked not only on the theory of symmetry groups, but I also had pieces of work on nuclear physics and even on the theory of accelerators, but, nevertheless, the theory of symmetry groups has always been my favorite subject. And certainly, my main work was connected with the application of this theory to the physics of elementary particles.

By main, I mean my work on the establishment of parastatistics, work on the discovery of the conspiracy of Regge poles, and on the application of symmetry groups to the classification of hadronic resonances, that is, to baryonic and mesonic resonances. But my most important result, which is the most widely-known one now in the world, is certainly the discovery[1] of a new symmetry group – supersymmetry – and the extension[2] of this to aspects of the general theory of relativity, the so-called supergravity. I can speak about this in some more detail. First, a couple of words about how I came to the discovery of supersymmetry and supergravity. The starting point was the idea of W. Heisenberg. Let me tell you what this idea was. At the end of the 60's, Heisenberg proposed the idea that all elementary particles can be described by a unified theory on the basis of the nonlinear equation formulated by him.[3] He started this work together with W. Pauli, but sometime later Pauli dissociated himself from this theory and even sharply criticized Heisenberg for his assumptions. But, nevertheless what did Heisenberg contribute to the discovery of supersymmetry? Heisenberg tried to explain the spectrum of all elementary particles on the basis of his theory. And thanks to his great intuition, he guessed the places of the particles and predicted how the photon should appear in this theory. Moreover, he found a place for the neutrino within his theory and assumed that the neutrino emerges as a result of spontaneous symmetry breaking. This was a very unusual assumption, because all known Goldstone particles emerging as a result of spontaneous symmetry breaking had spin zero in all theories known at that time. This idea of Heisenberg was revolutionary, because he was the first to formulate that the thought that there might exist Goldstone particles in nature with spin one-half. To tell the truth, he found this particle by an incorrect method, but nonetheless it was an idea that had a strong impact on me and from that time I often, although not constantly, would think about whether this idea could be realized. And when I started to consider how such particle might appear in the theory, I understood that this requires an extension of the usual physical groups, which is the basis for all relativistic processes. That means an extension of the Lorenz group and an extension of the Poincaré group so that new operators would be present, which would correspond to a quantum number of the neutrino. Thus, the main result that we obtained was that we managed to create such an extension of the Poincaré group, which is now called the Poincaré supergroup.

Later on, we encountered certain difficulties and, if one speaks about the direct application to the neutrino, this idea nonetheless did not work. Why? Because the group that we were considering contained the Poincaré group and it is known that the Poincaré group leads eventually to the general theory of relativity, if one considers the local transformations of the group. That means that the same local transformations of the supergroup would lead to the emergence of certain superpartners of the ordinary gravity field, and these superpartners would totally absorb the Goldstone particles. And thus, when I understood such ideas, I, together with my collaborators, proposed an extension even of Einstein's general theory of relativity, which would include just what we call supersymmetry. So, in fact, spin 3/2 superpartners would emerge. Moreover, in the extension of this theory not only particles with spin 3/2 emerge, but also particles with spin 1, with spin 1/2 and with spin zero. That is, there arose the idea that all elementary particles might be included into the system of supergravity. Now many scientists all over the world are working on it. Different variants of supergravity are being considered and also different variants of superstring theory, which are based on a certain extension of supergravity that takes into account the fact that elementary particles can be not simply point objects but also extended objects. This direction is now the main direction in theoretical physics and proposes that there is a unified theory of all elementary particles, based on one common principle – namely, on local supersymmetry.

Heisenberg's idea was, if one can say so, a physical idea, but in order to give this idea shape in a new precise mathematical theory, a certain mathematical formalism was required. And here I am very thankful to J. Schwinger, whom I personally knew, and who in a number of cases could have discovered supersymmetry. Moreover, I have recently seen an article by Schwinger, in which he writes why he didn't discover supersymmetry in spite of the fact that he was completely ready for it.

Why did I think that Schwinger could have discovered supersymmetry? Firstly, he introduced the concept of certain anticommuting variables, which played the roles of physical variables in field theory. He was the first to do this. Secondly, when I was saying that in supergravity the superpartner of the gravity field appears, well, these superpartners are particles of spin 3/2, and the theory of such spin 3/2 particles was first developed by Schwinger. It is even called the Rarita-Schwinger theory.

Recently, I received this article by Schwinger in which he writes about the reasons why he did not discover supersymmetry. Schwinger also says that he was very close to the discovery of supersymmetry and explains why. At the same time, in response to a general question as to why he didn't do it, which he received from the audience in the auditorium, he answered that this is a philosophical question and that this philosophy of discoveries is discussed in a number of monographs and that he could give only a general answer to the question.

The next moment that played an immense role in the mathematical formulation of supersymmetry, this was the work of the greatest French mathematician Elie Cartan. He created a special formalism of differential geometry, which seem to be specially appropriate for the system that I was considering. This particular circumstance, that I knew this work very well, helped me to create the mathematical formalism of supersymmetry. So, I think that there were actually three sources that helped me to develop the theory of supersymmetry and the theory of supergravity: the idea of Heisenberg, the idea of Schwinger and the mathematical idea of the great mathematician Elie Cartan.

Q. Dmitry Vasilyevich, I am not a theoretician, and it is very difficult for me to assess your accomplishments. Could you tell me in popular terms how your work is recognized and about its place in theoretical science.

A. It's very easy to answer this question, mostly because there are many references to our work. The number of references is more than a thousand. This is a very large indicator for scientific works. Besides that, testimony to the fact that this work is recognized in the world is the fact that in 1994 I gave a talk at a very important conference where the results of scientific development for the past 50 years were summarized and where the speakers were authors who had made fundamental contributions to the development both of theoretical ideas and in the acquisition of new experimental data.

Q. Where was this conference?

A. It was in Erice, in Italy. It was called the "International Conference on the History of Original Ideas and Basic Discoveries in Particle Physics." I was invited to speak on supergravity, and it was a great honor for me. And so I gave my talk at this conference.[4]

Q. Dmitry Vasilyevich, who else was studying the same problems that you were working on at that time?

A. At that time, in fact there was one more work that was done at the same time as ours. This was the work by Yu. Golfand and E. Likhtman, who introduced formulations of the theory of supersymmetry, but for completely different reasons. They tried to explain on the basis of supersymmetry the breaking of the parity that exists in nature. Unfortunately, Golfand passed away in 1994. There was no further work. If one talks about supergravity, our first work was done in 1973. The next work in this field appeared in the West only in 1976.

Q. Dmitry Vasilyevich, could you please tell us about your scientific contacts with your colleagues?

A. At KhFTI, when I was starting my work, there was a group studying quantum electrodynamics. Later on, I worked practically alone. But I had very good contacts with physicists from Moscow and Leningrad. I would like in particular to stress the role of Isaac Yakovlevich Pomeranchuk, who was working at that time in Moscow.

Q. Did you know him personally?

A. I knew him personally. Every time I came to Moscow, I always visited Pomeranchuk. We had lengthy discussions together and it was interesting that we had the same way of thinking, because we usually discussed the conceptual part of work without going into formulas. We didn't even write formulas, but clarified what connections are possible between phenomena. I would like to tell about the following episode as an interesting example. It was in 1962 and I had been sent as a member of a delegation from the Soviet Union to CERN. I was lucky to share a hotel room with Isaac Yakovlevich Pomeranchuk. This was a stressful time for me, because just then I had finished some work together with V.N. Gribov about the conspiracy of Regge poles, which I spoke about earlier. This was the most important subject for me at that time. So, I remember lying on a hotel bed and that I couldn't fall asleep. It was already two o'clock in the morning and a certain idea came to me. I shook Isaac Yakovlevich awake, he awoke and said "What!" and I told him about this idea. After that, we discussed this idea for a couple of hours without turning on the light and we fell asleep only at dawn. This certainly characterizes a person, if he can discuss a topic that interests him at any time of day or night. Such a person was Isaac Yakovlevich Pomeranchuk. [...]

Now, about Kharkov. At first I worked in relative solitude. Then I managed to form a small group of young capable scientists. The personnel of my laboratory consist of about eight people, and I continue working with them.

References

1. D.V. Volkov and V.P. Akulov, *Phys. Lett.* **B46**, 109 (1973).
2. D.V. Volkov and V. Soroka, *JETP Lett.* **18**, 312 (1973).
3. W. Heisenberg, *Introduction to the Unified Field Theory of Elementary Particles*, (Interscience Publishers, London 1966).
4. D.V. Volkov, *Supergravity Before 1976*, in *History of Original Ideas and Basic Discoveries in Particle Physics*, Eds. Harvey B. Newman and Thomas Ypsilantis, (Plenum Press, New York, 1996) [hep-th/9410024].

NONLINEAR WAY TO SUPERSYMMETRY

V. AKULOV[a]

Baruch College, CUNY, 17 Lexington Ave, New York, NY 10010

This year the physics community will celebrate the 30-th anniversary of supersymmetry. I want to tell about the first steps in this direction made by our group headed by Professor Dmitri Volkov, in Kharkov.

Supersymmetry is an extremely interesting mathematical structure which arose in high energy physics in the early seventies independently in Moscow (Golfand and Likhtman), in Kharkov (Volkov and Akulov), and CERN (Wess and Zumino). Supersymmetry differs from previous symmetries in that it unifies (i) bosonic and fermionic particles as members of one supermultiplet, and (ii) the space-time Poincaré symmetries with the internal (isospin) symmetries.

The first published mention of the super-Poincaré algebra in four dimensions was in 1970. In this year the International Rochester conference on High Energy Physics was held in Kiev, Ukraine. Prof. Yu. Golfand announced two reports for this conference, but the organizing committee gave him time only for one report, so he preferred to report on the vacuum in QED. The problem of the super-Poincaré algebra which had been already obtained by him at that time seemed too complicated to be presented at the conference. A collection of abstracts of all talks submitted to the Conference was published by the Institute of Theoretical Physics in Kiev, as a rotaprint edition (edited by L. Yenkovsky *et al*). The first volume carried a picture of Taras Shevchenko on the cover, the second volume the picture of Bogdan Khmelnitsky. Golfand's abstract of the report on the super-Poincaré algebra was there. Yu. Golfand, together with his Ph.D. student E.P. Likhtman, found this algebra trying to unify the parity transformation with the Poincaré group. Unfortunately, now I cannot find this rotaprint edition of the collection of abstracts, while the final volume of the Proceedings did not contain any mention of those talks which had not been delivered.

---

[a]e-mail address: akulov@gursey.baruch.cuny.edu

The mathematical background for supersymmetry had been constructed even earlier: in 1970 Felix Berezin (Moscow) and Gregory Katz (Kiev) published[b] a paper on groups with commuting and anticommuting parameters. In this way the supergroups appeared in mathematics.

In the 1950's and '60's Dmitri Volkov investigated the connection between spin and statistics and rediscovered (after M. Green, 1953) the parastatistics in 1959. Later he considered fermionic Regge trajectories. The success of the application of the Goldstone theorem to π meson physics stimulated the desire to generalize this theorem to include fermions. Another hint was linked with Heisenberg's (incorrect) idea that the neutrino is possibly a Goldstone particle connected with a (spontaneously broken) discrete symmetry – P parity.

I was a graduate student at this time, and Professor Volkov suggested this topic for my Ph.D. work in 1971.

Using the Berezin-Katz formalism we constructed an extension of the Poincaré group which included fermionic supercharges with anticommuting parameters and $U(n)$ – the internal group, which is now called the extended Poincaré supergroup. The parameters of this supergroup form a superspace (introduced later by A. Salam).

At first we constructed the exponential representation of the super-Poincaré group, but this approach was very complicated. The last part of the paper by Berezin and Katz contained an example of the matrix realization of supergroup, based on the two-by-two Pauli matrices. We constructed a matrix representation for a supergroup using three-by-three and four-by-four matrices. Then, using the well-known representation for the Poincaré group as the upper triangular matrix, we constructed an extended Poincaré supergroup, which included translations of the spinor Grassmann coordinates and a unitary group of the internal symmetry. The coset space of this supergroup included the Grassmannian coordinates with the unitary index along with the conventional (four-dimensional) Minkowski space.

Basing on the extended Poincaré supergroup, we constructed a nontrivial unification of space-time symmetries, like the Lorenz group or the Poincaré group, with the internal symmetries. In this way we

---

[b]F. Berezin and G. Katz, *Mathematicheskiy Sbornik*, **83** 343 (1970) [*Math. USSR – Sbornik*, **11** 311 (1970)].

bypassed the no-go theorem of Coleman and Mandula.

The above coset was later called *Superspace* by Salam and Strathdee. Our next step was the construction of the action integral invariant under such group, that would describe a spontaneously broken supersymmetry. (Of course, we did not use the word *supersymmetry*, it was put in circulation later.) Thus, a general theory of the spontaneously broken supersymmetry emerged. Our action had a nonlinearly realized supersymmetry. It had the form of the Born-Infeld action in the absence of the gauge F field. As was noted by Renata Kallosh in 1997, from the modern perspective one can say that it was one of the first D3 branes.

During the summer of 1972 we finished this work, and sent the paper to JETP Letters and to Physics Letters. The latter was a great problem for our Institute, because all paper intended for publication abroad had to be cleared in Moscow. This would normally take 3 months or more. Only after a positive decision from Moscow could we send our papers abroad.

In the autumn of 1972 we attended the International Seminar in FIAN, Moscow, on the $\mu - e$ problem. Professor Volkov reported on the possible universal neutrino interaction. Efim Fradkin invited him to give a two-hour talk at the Theoretical Division of FIAN. At that time Professor Victor Isaakovich Ogievetsky told us about the paper by Yu. Golfand and E. Likhtman in JETP Letters, that had a similar algebra. Yu. Golfand (who worked in FIAN) was absent during Volkov's talk. Likhtman also did not attend it.

We returned to Kharkov and read the paper by Golfand and Likhtman. This paper contained the action of an Abelian gauge model with the linearly realized supersymmetry. We added the reference to their paper, and sent our detailed report to Moscow, to *Theoreticheskaya i Matematicheskaya Fizika*. This detailed paper contained supergroups as matrix constructions, and the interaction of the Goldstone fermions with other fields.

After global supersymmetry D.V. Volkov considered a local version, and in 1973 (together with V.A. Soroka, because I finished my Ph.D work then) constructed the first version of supergravity, which consisted of graviton with spin 2, gravitino with spin 3/2 and a Goldstone fermion with spin 1/2. But the breakthrough began only after the paper by J. Wess and B. Zumino in 1973, but this is another story.

FROM SYMMETRY TO SUPERSYMMETRY [a]

JULIUS WESS

Sektion Physik der Ludwig-Maximilians-Universität
Theresienstr. 37, D-80333 München

and

Max-Planck-Institut für Physik
(Werner-Heisenberg-Institut)
Föhringer Ring 6, D-80805 München

"...die Wirklichkeit zeigt sich in unseren Erlebnissen und Forschungen nie anders wie durch ein Glas, das teils den Blick durchläßt, teils den Hinein-blickenden widerspiegelt."

R. Musil "Der Mann ohne Eigenschaften",

Zweiter Teil, Kapitel "Das Sternbild der Geschwister oder Die Ungetrennten und Nichtvereinten"

Thank you for the invitation and your kind introduction. It is a pleasure to be here, and this for many reasons. Let me mention one explicitly. This is the place where Golfand worked during his last years as a member of this faculty and I would like to dedicate this lecture to his memory. As you know, he was one of the founders of supersymmetry.

It is quite easy to speak about symmetries. Everybody has a notion of symmetry, it is a very deeply rooted and widespread concept, ranging from art to science. In some way or other symmetry is perceived by everybody. I think it is worth mentioning that about thirty years ago there was strong interest in experimenting with apes to see how much they were able to learn.[1] One objective was to see how apes would learn to paint. In one of these experiments one dot was made at one side of a piece of paper and the ape would then try

---

[a] This is a write-up of the first lecture in the 1999 Technion Distinguished Lecture Series, delivered on March 10, 1999. The draft of the write-up was prepared by Dmitry Gangardt from the tape recording. The lecture was first published in the Yuri Golfand Memorial Volume *The Many Faces of the Superworld* (World Scientific, Singapore, 2000), page 85.

to make a dot on the other side to balance it symmetrically. That's exactly what we are doing in physics.

It is quite difficult to speak about symmetries. Everybody has his own concept of it, also physicists, and you never know if we communicate about the same thing.

Fortunately, mathematics with its strong capability to abstract has abstracted the concept of symmetries to the concept of groups. Group theory and the theory of representations of groups incorporate many of the different aspects of symmetries as we meet them in nature, in art, in science, etc. When referring to symmetry I mean it in the framework of group theory, representation theory, algebra and differential geometry.

In physics symmetries have been used all along. Group theoretical methods make it much simpler to get information about a system. A problem like the Kepler problem, for example, would be much harder to solve without using rotational symmetry.

Through symmetries you might get information about a system without really understanding the physical laws that govern it or without being able to solve the dynamical laws if you know them. You may think about the trivial example of a scale. Without knowing the theory of gravitation you are quite convinced that, based on the symmetry of a scale, the weight on the left hand side and on the right hand side of the scale is the same. This demonstrates how symmetries can help us to find our way through a system without really knowing its laws.

In addition symmetries have a very strong interplay with experimental physics via conservation laws. Conservation of energy, momentum and angular momentum can be measured experimentally. They are linked to an invariance under time translation, space translation and space rotation. Nœther's theorem states it very precisely: if we know that a system is invariant under some symmetry transformations, then we can show that there are corresponding conservation laws and we know how to find their explicit form. We know that there is a conservation law of the electric charge — we ask for a symmetry and find it in the phase transformation of the Schrödinger wave function. This shows that there is a very strong connection between abstract mathematical concepts and experimental facts. We are very lucky to have such an interplay in physics.

Symmetries in modern physics have taken an even stronger role to such an extent that the laws of modern physics cannot even be formulated without the concept of symmetries. To make the framework of local quantum field theory meaningful, symmetries have to be invoked from the very beginning. It is not that we know the laws and try to find their symmetries, but rather we have to implement the symmetries from the very beginning to be able to formulate these laws in a meaningful way.

I have prepared a transparency listing some of the fundamental symmetries (Fig. 1). It shows two separate columns: one, on the left, for space-time symmetries and one, on the right, for symmetries in an inner space.

At the top of the left column we have the rotation group. It is the group from which a physicist can learn what group theory is about. We know that the laws of electrostatics and magnetostatics are invariant under the rotation group and that Newton's laws allow rotational invariance as well. The rotation group is strongly related to $SU(2)$, the special unitary group in two dimensions, which is connected to the concept of spin. You know all this from quantum mechanics. When electrostatics and magnetostatics are combined to Maxwell's theory of electromagnetism we encounter the Lorentz group. Though electrostatics and magnetostatics describe forces that are different in strength by the order of magnitude of the velocity of light, they nevertheless are part of the same theory. We could say that the unification comes about by generalizing the rotation group to the Lorentz group. This is a good model and yields a good understanding of how theories can be unified by enlarging a group.

The right column shows symmetries in an inner space. I have mentioned the phase transformation of the wave function. Wave functions can span a higher-dimensional space and this is the space which we call inner space. The first step was done by Heisenberg. He had learned what $SU(2)$ is in the framework of the rotation group and spin and put this concept to work in the inner space as isospin. This was later generalized to $SU(3)$, a successful attempt, which we now understand in terms of the quark model. A very successful model, the Standard Model, is based on a group $SU(3)_C$ for color,

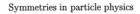

Symmetries in particle physics

Space-time	Inner space
Translations	Phase transformations
Rotations $SU(2)$	Isospin $SU(2)$
	$SU(3)$
Lorentz group	$SU(3)_c \times SU(2)_W \times U(1)$

gauging | Conformal group | GUT $SU(5)$ | gauging

Gravity | Standard Model
long distances | short distances 10^{-16} cm
Classiical theory | Quantum Field Theory

Figure 1: Symmetries in particle physics

$SU(2)_W$ for weak interactions and $U(1)$. The Lorentz group together with this $SU(3)_C \times SU(2)_W \times U(1)$ is at the basis of the Standard Model. This model describes physics very well as we know it in our laboratories, down to a scale of 10^{-16}cm.

It is natural to ask if the $SU(3)_C \times SU(2)_W \times U(1)$ symmetry is not part of a larger group, $SU(5)$, for instance. This then gives rise to a theory where strong, weak and electromagnetic interactions would be truly unified. Such a theory we call GUT — for Grand Unified Theory — but what Nature knows about this, we do not yet know. It is the Standard Model based on the Lorentz group and the group $SU(3)_C \times SU(2)_W \times U(1)$ that is supported and by now well tested by experiment.

Looking at the two columns it seems that Nature (or we) has used the same concept of symmetry twice. But has Nature chosen it separately? Has it done the same thing twice? It is natural to ask for a bridge between the two columns.

We have learned from the Maxwell equations that one can make a transformation parameter space-time dependent. This is the property of the gauge transformation in Maxwell theory. We can identify this parameter in Maxwell theory with the parameter of the phase transformation of the Schrödinger wave function and build a gauge theory, as we call it today. Thus the idea of a gauge theory is born. You ask that the theory should be invariant under a group that acts in inner space and has space-time dependent parameters. This is the concept on which the Standard Model is built. The Standard Model is the gauge theory for the group $SU(3)_C \times SU(2)_W \times U(1)$. If the same idea of gauging is used for the space-time symmetry, the Lorentz group, one arrives at Einstein's theory of gravitation.

We know that the Standard Model can be interpreted — via the concept of renormalization — as a Quantum Field Theory and, as such, it is experimentally extremely successful at short distances, the 10^{-16} cm I have mentioned before. We would have liked to find deviations from the Standard Model experimentally to get a hint where to go next: $SU(3)_C \times SU(2)_W \times U(1)$ seems to be the simplest choice one can make, and it works, and it works and all the experiments verify it again and again. Is Nature not more sophisticated? Doesn't it know about GUT?

As for Einstein's gravitational theory we know that it is a very good theory for long distances. We have no reason to doubt its validity there. Our understanding of space-time at large distances is based on this theory.

We have the situation that the Standard Model and Einstein's theory of gravitation describe the data observed in the laboratory as well as in astronomy and astrophysics very well. Even a cosmology based on these two theories is very reasonable.

Looking at our columns we now have an even more puzzling situation. The same concept of symmetries and gauging them gives, in the left column, a very good classical theory of gravity, defending itself against quantization by an abundant number of singularities. There seems to be a deep conflict between the classical theory of gravitation and quantum field theory. On the other hand the same idea about symmetry and gauging in the right column leads to a very good model of a renormalizable quantum field theory which is mathematically and experimentally successful.

Is gauging all or is there a deeper connection between space-time and inner space symmetries? In nuclear physics, a long time ago, Wigner and Hund[2] proposed a group $SU(4)$ that has $SU(2)$ of spin and $SU(2)$ of isospin as a subgroup. This way they unified space-time and inner space and got a good classification of nuclear levels. At the time of $SU(3)$ in particle physics, this idea was generalized to $SU(6)$, incorporating $SU(2)$ of spin and $SU(3)$ of inner space. Reasonable experimental predictions about masses of particles with different spin — spin 0 and spin 1 as well as spin $1/2$ and spin $3/2$ — can be based on $SU(6)$. This is now better understood on the basis of the quark model. As the quark model was not yet known at the time, many attempts were made to extend the $SU(6)$ model to incorporate the Lorentz group. But this proved to be impossible. It was impossible to build a Lorentz invariant model that at low energies would have $SU(4)$ or $SU(6)$ as a symmetry. If a lot of physicists try something and it does not work, then it might be clever to try to prove that it cannot work. And this is what was done. It started with work by O'Raifeartaigh[3] and came to a very elegant formulation which now is known as the Coleman – Mandula "no-go" theorem.[4] This theorem tells us that for a theory in four dimensions, with the Lorentz group as a symmetry group, and satisfying a certain number of axioms, which I am going to tell you about in a minute, the only possibility for a symmetry group is the direct product of the Lorentz group with some compact inner group. Surprisingly enough, we have a theorem that separates the two columns on the basis of very fundamental axioms.

Now some remarks on the axioms. Apart from the Lorentz invariance, the axioms state that it should be a theory based on quantum mechanics and that it should be local, i.e. it should be a local quantum field theory. Local here means local in the microscopic sense. Two measurements that are separated by a space-like interval cannot influence each other, no matter how small the distance would be. The locality of the theory is based on the locality of the fields. In addition we assume that there is a unique lowest energy state, a ground state which we call vacuum, and that all the other states have larger energy. Probability is supposed to be conserved in the quantum mechanical sense. Finally we assume that there is only a finite number of different particles. These seem altogether very reasonable assumptions, but they have as a consequence that you cannot combine the two columns, as we tried by postulating a $SU(4)$ or a $SU(6)$ symmetry.

There is one problem with this set of axioms. With the exception of the free field theory we do not know a local quantum field theory the existence of which we can rigorously prove and which satisfies all the axioms. We have invented very powerful methods of perturbation theory, of separating infinities, going through a renormalization scheme. We are able to extract in this way information that we can test experimentally. In some way it is a kind of art, but it works beautifully. On this basis we understand the models I have been talking about and we make them successful. This is, however, a formulation of a model a mathematician would not like to accept as a theory. But the success in comparing it with experimental data is so strong that we cannot dismiss this type of theories.

Now I would like to make a point. If we try to build a theory on the setting just discussed and we want to have spin-zero, spin-one and spin-one-half fields, then we start from a free field theory and try to arrange the multiplets and couplings such that there should be only a finite number of divergences in perturbation theory. We start with the tree level, study the energy-momentum dependence of the Feynman diagrams and arrange the model in such a way that this dependence is as smooth as possible in order to facilitate integration over energy and momentum variables at the one loop level. The result is a gauge theory with spontaneous symmetry breaking. Thus, without knowing about group theory and symmetries the whole concept of gauge theories could be invented by a good physicist on dynamical grounds. This was shown about twenty years ago by John Cornwall *et al.*, Llewellyn Smith and Steve Weinberg.[5] They pioneered this approach. After you know this result it is much easier to take a textbook on group theory, formulate your model based on your knowledge of group theory, and to gauge the symmetry group. In this way you obtain a model and you will find it to be a renormalizable quantum field theory. This can be proved from gauge invariance. Nœther's theorem and the conserved currents are at the heart of this proof. The current has to be a well-defined object and it helps to relate infinite and, therefore, undefined constants such that there is only a finite number left at the end. This defines the renormalization scheme. This is what I meant in the beginning when I said that even in formulating a physical theory we have to use the concept of symmetry. But this also raises the question whether symmetries are very basic or whether they can be derived from the dynamics of the system – based on some reasonable axioms.

Our way of thinking is very much influenced by the deep and widespread notion of symmetry to an extent that if we find a beautiful symmetry in a dynamical system we say that we understand this system. If there is a deviation from the symmetry we say: there must be something going on which we do not understand.

There is another good reason to start from symmetries instead of dynamical requirements. The second way would be difficult and, let me say, ugly. Mathematicians have not developed concepts along this ill-defined line. Thus, we have no mathematical machinery that we can use, contrary to the first approach, where we have all this beautiful mathematics.

There is a surprising way of combining space-time symmetries with internal symmetries, and this is by supersymmetry. This is achieved by generalizing the concept of symmetry. As you know, symmetries can be formulated in terms of commutator relations, as the commutator relations of angular momentum in quantum mechanics. A large class of groups – the Lie groups – are related to Lie algebras that are defined by commutator relations. All the symmetries we have mentioned are of this type.

From Dirac we know that not only commutators but also anti-commutators are a very useful concept, especially when we deal with particles with half-integer spin.

The idea is to generalize the concept of symmetries to a structure which is formulated in terms of commutators and anticommutators as well. This is not a new idea in mathematics, such structures – graded Lie algebras — have been thoroughly investigated e.g. by Berezin.[6] But can such a concept be realized in a quantum field theory? The answer is yes. There is a quite unique symmetry and its uniqueness is based on the no-go theorem of Coleman and Mandula (Fig. 2). This was shown by Haag, Lopuszański and Sohnius.[7] Here the theorem does not tell you that it does not go, it tells you that it goes in a very unique way expressed by the formula:

$$\{\bar{Q}^N{}_\alpha, Q^M{}_\beta\}_+ = 2\gamma^\nu{}_{\alpha\beta} P_\nu \delta^{NM}. \tag{1}$$

The charges $Q_\alpha{}^N$ are spinorial charges, α is a spinor index and $Q_\alpha{}^N$ is a Majorana spinor, N is a free index. If N goes from one to two we speak of $\mathcal{N} = 2$ supersymmetry, if it goes from one to four we speak of $\mathcal{N} = 4$ supersymmetry. The four vector P^ν is the energy-momentum four-vector that generates the translations in space-time. The structure constants of this algebra are the Dirac γ matrices and the Kronecker δ-symbol. This algebra can be combined with the algebra of the Lorentz group, then it is the algebra of supersymmetry.

These spinorial charges can be realized in terms of local fields, and the algebraic relations of supersymmetry hold on the basis of the canonical commutation relations of the fields. The charges are related to currents, which are spin-3/2 currents, in the same way as the energy-momentum vector is related to the energy-momentum density tensor which is a spin-2 object. If the theory is gauged, the currents are the sources of fields; for the electric current this is the photon, with spin one, for the energy momentum tensor it is the

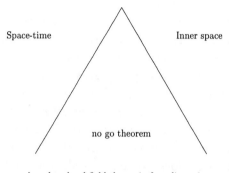

Space-time   Inner space

no go theorem

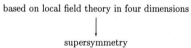

based on local field theory in four dimensions

supersymmetry

Figure 2: Superextension of the Poincaré algebra

graviton, with spin two, and for the supercurrent it is the gravitino with spin 3/2.

Supersymmetric theories have a very encouraging property. As quantum field theories they are less divergent as they would be without supersymmetry.

Supersymmetric theories have a very discouraging property. It follows directly from the algebra that a supersymmetric theory has to have an equal number of bosonic and fermionic degrees of freedom degenerated in mass. This is not how Nature is. We have fermions and bosons but not in the same multiplet structure and in no way degenerated in mass.

The two properties lead to frustration, the more so as they are not independent. Let me explain this with a simple example. Take a bosonic and a fermionic harmonic oscillator, both with the same frequency. The zero point energy will have opposite sign and adding them leads to the cancelation of the zero point energy. A field can be viewed as an infinite sum of harmonic oscillators. The zero point energy will add up to an infinite vacuum energy, except in a theory where the bosonic and fermionic contributions cancel. This was already known to Pauli but he also knew that the world is not like this. Supersymmetry relates that cancelation to an algebraic structure of the theory, and you might be led to believe that it is now based on a deeper property of the theory and nature. Since the days of Pauli we have learned to deal with symmetries that are spontaneously broken. The field-theoretic properties of such theories with spontaneously broken symmetries are maintained, but at a phenomenological level at low energies the symmetry appears to be broken by sizable effects.

The same mechanism that leads to the cancelation of the vacuum energy leads to many other cancelations of divergences. These improved renormalization properties of the theory can be traced back to the cancelation properties of diagrams with bosonic or fermionic internal lines. It can be shown that there are parameters in the theory which do not get any radiative correction, not even a finite one. If not a miracle, this is a sensation in quantum field theory. You can introduce parameters like certain masses or couplings that are not changed by radiative corrections at all. Naturally, this has consequences for particle phenomenology based on supersymmetric theories.

Let me first discuss the fact that supersymmetric theories have an equal number of bosons and fermions.

In Nature we know quarks and leptons to be fermions. These are the particles that constitute matter. Each of these fermions has to have two bosonic partners as each spin-one-half particle has two degrees of freedom. We do not know such partners in Nature but we can give them names. The SUSY partners of the quarks we call squarks and the SUSY partners of the leptons we call sleptons. The "s" stands for scalar of supersymmetric partner.

In Nature we know the photon, the vector particles of weak interaction, the gluons, the graviton and the Higgs to be bosons. These are the particles that constitute the forces. Each of these bosons has to have a fermionic partner. We give them names: photino, wino, zino, gluino, gravitino and Higgsino.

With these particles we can build models and consider it a success that we know already half of the particles in such models. But we know also their couplings, which are entirely determined by the couplings of the particles we know. Take a Feynman diagram with the known particles (Fig. 3). A line in such a Feynman diagram that either goes from an incoming particle to an outgoing particle or that forms a closed loop can and has to appear in an equivalent diagram where it is replaced by a line that is associated with SUSY partners. Doing this for all the lines you obtain all the diagrams of a certain supersymmetric theory. The coupling constants on the respective vertices are the same as in the theory you started from.

In a truely supersymmetric theory the masses would also be the same. Not so if the symmetry is spontaneously broken. We know that for spontaneously broken gauge symmetries the mass difference within a multiplet might be quite big — like the mass difference between the photon and the W or Z.

It is possible to break supersymmetry as well in such a way that its renormalization properties remain valid, the mass difference between SUSY partners can then be quite big — we call it the SUSY gap.

The radiative corrections, however, will not completely cancel — we can expect finite contributions which then will also be of the

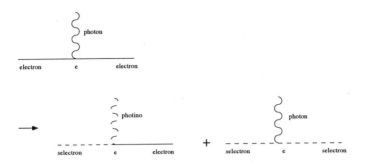

Figure 3: Supersymmetric vertices in QED

order of the SUSY gap. Among the masses and coupling constants of a gauge theory that will have this property are the Higgs masses and the Higgs couplings.

Knowing this one can relate this property to another phenomenon in particle theory. Assume that there is a GUT theory. The unifying gauge group has to be spontaneously broken to render the Standard Model as we know it. The symmetry group $SU(3)_C \times SU(2)_W \times U(1)$ of the standard model is then spontaneously broken to the theory of electromagnetism and weak interactions as we know it at low energies.

This breaking of the symmetries is triggered in a quantum field theoretical model by parameters, the Higgs masses and Higgs couplings, which have to be renormalized in a non-supersymmetric theory by an infinite amount. It is very difficult to understand that the breaking scheme in the two sectors of the breaking is stable and respects the scale. This is the hierarchy problem in particle physics. If, however, the theory is a spontaneously broken supersymmetric theory, then the relevant parameters obtain only radiative corrections of the order of the SUSY gap. If the SUSY gap is of the order of the electroweak breaking scale characterized by the W mass then we would understand the stability of the breaking of the symmetries in a GUT theory. In such a scenario the SUSY gap has to be of the order of one TeV – an order which will be accessible to experiments soon.

My personal belief is that it would be a waste having such a beautiful symmetry as supersymmetry just to solve the hierarchy problem. Supersymmetry might play a much more fundamental role at higher energies and solving the hierarchy problem would just be one of the lower energy remains of supersymmetry.

But let me stress that there are two inputs in this prediction – SUSY and GUT. In any combination they might be right or wrong.

Now back to the theme mentioned already in the context of gauge theories. We see that supersymmetry has a very strong influence on the dynamical behavior of quantum field theory when it is treated in the framework of renormalization theory. We can start from this framework and ask for a model, e.g. with spin 0 and spin 1/2 particles, that behaves as smoothly as possible not only at the tree level but on the one-loop level as well. We could ask for the cancelation of the infinities in the vacuum energy, in this way we would establish an equal number of fermionic and bosonic degrees of freedom. If we then asked for the non-renormalization properties about which I have been talking before we would construct a supersymmetric theory — with, what we call possible soft breaking, a symmetry breaking that does not affect the renormalization behavior. If we asked for the absence of any radiative correction we would obtain a strictly supersymmetric theory.

So again it would be a dynamical concept, now put forward to the level of one-loop diagrams which would have led to the invention of supersymmetry, without knowing about any algebra, about groups or graded groups, just being a good physicist knowing how to handle the Feynman diagrams. Somehow supersymmetry is the next logical step after gauge theory in the framework of renormalizable quantum field theories. You go from the tree level corresponding to a classical theory to the quantized level represented by one loop, apply the same idea once more, and you arrive at supersymmetry. What we do not know is if Nature knows about this way of thinking or if Nature has a different logic, does things differently.

Now to the history of supersymmetry. It started with the work of Golfand and Likhtman.[8] They thought about adding spinorial generators to the Poincaré algebra, in that way enlarging the algebra. This was about 1970 and they were really on the track of supersymmetry.

I will come back to this idea at the end of this talk as I think that this is the right question: can we enlarge the algebra, the concept of symmetry, by new algebraic concepts in order to get new types of symmetries?

Then in 1972 there was a paper by Volkov and Akulov[9] which argued on the following line. We know that with spontaneously broken symmetries there are Goldstone particles supposed to be massless. In Nature we know spin one half particles that have, if any, a very small mass, these are the neutrinos. Could these fermions be Goldstone particles of a broken symmetry? Volkov and Akulov constructed a Lagrangian, a non-linear one, that turned out to be supersymmetric. Of course today we know from Haag, Lopuszański and Sohnius that it had to be supersymmetric. But being nonlinear, just as the nonlinear sigma model, the Lagrangian is highly non-renormalizable and does not show any sign of the renormalization properties which we now find so useful and intriguing in supersymmetry.

Another path to supersymmetry came from two-dimensional dual models. Neveu and Schwarz et al.[10] had constructed models which had spinorial currents related to supergauge transformations that transform scalar fields into spinor fields. The algebra of the transformation, however, only closed on mass shell. The spinorial currents were called supercurrents and that is where the name "supersymmetry" comes from.

In 1973 Bruno Zumino and I published a paper[11] where we established supersymmetry in four dimensions, constructed renormalizable Lagrangians and exhibited non-renormalization properties at the one-loop level. Our starting points were the supercurrent and the strong belief in Nœther's theorem.

Another paper by another author that could have led to supersymmetry based on the non-renormalization properties of perturbative quantum field theories was never written.

With supersymmetry it is very natural to extend the concept of space-time to the concept of superspace. Energy-momentum generates translations in four-dimensional space-time, so it is natural to have the anticommuting charges generate some translations in an anticommuting space. This new space together with the four-dimensional Minkowski space is called superspace.

Table 1:

spin:	0	1/2	1	3/2	2	
		2	1				ϕ, scalar multiplet
			1	1			V, vector multiplet
					1	1	gravitational multiplet

Fields will now be functions of the superspace variables and they will incorporate SUSY multiples in a very natural way. This idea was pioneered by Salam and Strathdee.[12] Lagrangians can be formulated very elegantly in terms of superfields and the non-renormalization theorems find very elegant formulations as well, as first shown by Fujikawa and Lang.[13]

The spin 0, spin 1/2 multiplets find themselves in the so-called chiral superfields (ϕ) and spin 1, spin 1/2 in the so-called vector superfields (V) (see Table 1).

Any diagram that has only external chiral fields (but no conjugate chiral field ϕ^* or vector field) will not be renormalized (Fig. 4). As no external field is in the class of chiral fields only we find the non-renormalization theorem for the vacuum energy. A short look at the superfield Lagrangian

$$\mathcal{L} \sim (\phi^* e^{gV} \phi)_D + (m\phi^2 + \lambda\phi^3)_F \tag{2}$$

tells us that the mass terms of the chiral fields are of the form ϕ^2 and the coupling terms of the Higgs type are of the form ϕ^3. These couplings do not get renormalized. The kinetic term of the chiral field is of the form $\phi^*\phi$, there is a wave function renormalization.

The superspace variables also play an important role if we want to gauge an internal symmetry of a supersymmetric theory. Gauging an internal symmetry means formulating a theory that is invari-

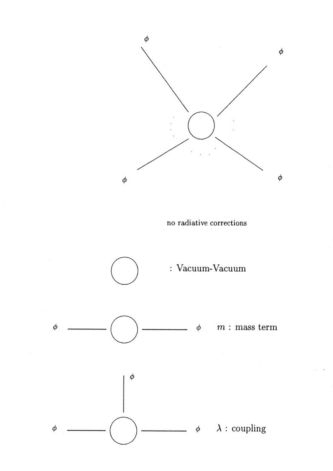

Figure 4: Non-renormalization

ant under transformations with space-time dependent parameters. Space-time by itself is not a supersymmetric concept, we have to replace it by superspace. Gauging an internal symmetry in a supersymmetric theory means to formulate a theory that is invariant under transformations with superspace dependent parameters. This way we know how to supersymmetrize all the known gauge theories.

Gauging supersymmetry as well finds a natural formulation in superspace. Einstein's theory of gravity can be formulated as a geometrical theory in four-dimensional space-time. Supergravity, the theory that has supersymmetry gauged, finds its formulation as a geometrical theory in superspace. Supergravity incorporates Einstein's theory because supersymmetry incorporates the Lorentz group. It improves the renormalization properties of the usual gravity theory, however, it does not make it a fully renormalizable theory. But it is closer to a dream to have also gravity as a part of renormalizable quantum field theory.

All our interplay with symmetries has centered around the renormalization problem of perturbative quantum field theory. The singularities that have to be renormalized are consequences of the unsatisfying short distance behavior of quantum field theories. Symmetries to some extent improve the situation. But is this really the way how Nature solves the problem of divergences at short distances?

Another possibility is to loosen some of the axioms. But this has to be done in a very controlled way, first so as not to get into conflict with experimental facts, and secondly in order not to have the rules of the theoretical game too wide open and to turn theory into a book-keeping device.

The only way up to now that meets these requirements is string theory. The concept of a point has changed to the concept of a string — the axiom of locality has been loosened and there are infinitely many particles, the excitations of the string in the theory.

At low energies the string picture might be compatible with our present experimental knowledge summarized in the Standard Model and the Einstein theory of gravity. The theoretical framework of strings is again based on symmetries and differential geometry. Analyticity plays an important role as well. In string theory our concept of space-time is based on differential manifolds — the closest relatives to flat space, however curved as they might be.

In the spirit of Golfand we could ask for a change in the algebraic structure of quantum field theory. We have based it on the canonical structure of quantum mechanics and adapted it to the algebraic possibilities of differentiable manifolds. Non-commutative differential geometry might give a mathematical frame that goes beyond differential manifolds. Attempts in this direction show that this could lead to a lattization of space-time at short distances. This opens possibilities worthwhile exploring. Maybe Golfand was at the track of another theoretically successful development again.

Acknowledgments

This article is based on a talk given at Technion, Haifa, in March 1999. I would like to thank Dmitry Gangardt who wrote the first draft on the basis of a tape recording. Especially I would like to thank Technion and Michael Marinov for their cordial hospitality.

References

1. D. Morris, *The Biology of Art*, (Methuen & Co.Ltd, London, 1962).
2. E. Wigner, *Phys. Rev.* **51**, 106-119 (1936);
   F. Hund,*Z. Phys.* **105**, 202-228 (1937).
3. L. O'Raifeartaigh, *Nucl. Phys.* B **96**, 331 (1975).
4. S. Coleman and J. Mandula, *Phys. Rev.* **159**, 1251 (1967).
5. J. M. Cornwall, D. N. Levin and G. Tiktopoulos, *Phys. Rev.* D **10**, 1145-1167 (1974); C. H. Llewellyn Smith, *Phys. Lett.* B **46**, 233 (1973);
   S. Weinberg, *Phys. Rev.* D **7**, 2887-2910 (1973).
6. F.A. Berezin, *The Method of Second Quantization* (Academic Press, New York, 1966); F. A. Berezin and G. I. Katz, *Lie groups with commuting and anti-commuting parameters*, in *Mat. Sbornik* (Engl. Transl.: *Math. USSR-Sb.*) **82**, 343 (1970).
7. R. Haag, J. Lopuszańsky and M. Sohnius, *Nucl. Phys.* B **88**, 257 (1975).

8. Yu.A. Golfand and E.P. Likhtman, *JETP Lett.* **13**, 323 (1971).

9. D.V. Volkov and V.P. Akulov, *JETP Lett.* **16**, 438 (1972).

10. A. Neveu, and J. H. Schwartz, *Nucl. Phys.* B **31**, 86 (1971).

11. J. Wess and B. Zumino, *Nucl. Phys.* B **70**, 39 (1974).

12. A. Salam and J. Strathdee, *Nucl. Phys.* B **76**, 477 (1974).

13. K. Fujikawa and W. Lang, *Nucl. Phys.* B **88**, 61 (1975).

3

Local Supersymmetry
(Supergravity)

D. Volkov

V. Soroka

S. Ferrara

D. Freedman

P. van Nieuwenhuizen

S. Deser

B. Zumino

THE SOURCES OF SUPERGRAVITY

VYACHESLAV A. SOROKA[a]
*Institute of Theoretical Physics, National Science Center "Kharkov
Institute of Physics and Technology," 310108 Kharkov, Ukraine*

Recollections on how the basic concepts and ingredients of supergravity were formulated by Dmitri V. Volkov and the present author in 1973-74.

When I was an undergraduate student at the Kharkov State University in 1965 my advisor Professor Dmitri Vasilyevich Volkov gave me a task to learn and to study the problem of quantization of the spin 3/2 field interacting with other fields. There was a problem with constructing selfconsistent theories for the interacting fields with high spins and to find the methods for quantization of such theories. In particular, Johnson and Sudarshan (1961) encountered the inconsistency under quantization of the spin 3/2 field minimally interacting with the external electromagnetic field. At that time I was a starting physicist and did not know the theories with high spins. So, I had to learn two formalisms for description of the spin 3/2 field: Bargman-Wigner and Rarita-Schwinger. By the way, from the paper of Rarita and Schwinger (1941) I learned and memorized for the future a remarkable fact that a free massless spin 3/2 field in their formalism possesses a gauge invariance with a spinor parameter. I also learned Schwinger's quantum dynamical principle within the framework of which Johnson and Sudarshan had tried to quantize the spin 3/2 field.

At that time we had no success in quantization of the spin 3/2 fields interacting with other fields. As became clear later on, the problem in that formulation had no solutions at all. But it was elucidated only after the discovery of supergravity theory in which the problem was naturally solved. So I was forced to postpone that approach and switched my activity to the derivation of sum rules following from the algebra of fields which had been introduced by Kroll, Lee, Weinberg and Zumino (1967). Since the algebra of fields is

---

[a]e-mail address: vsoroka@kipt.kharkov.ua

based on the gauge theory, I learned the beautiful Yang-Mills theory, and the papers by Utiyama (1956) and Kibble (1961) devoted to the gauge theories with different gauge groups, including, in particular, the gauge space-time symmetries connected with gravity.

Meanwhile Volkov (1969) simultaneously with Callan, Coleman, Wess and Zumino (1969) developed a general method for constructing Lagrangians for the Goldstone particles in the case of an arbitrary spontaneously broken internal symmetry group. Later he extended this method to the case of the spontaneously broken symmetries including the Poincaré group as a subgroup (1971, 1973).

Then Gol'fand and Likhtman (1971) introduced a super-Poincaré group in connection with a parity violation problem in quantum field theory. Shortly, and independently of them, Volkov formulated the existence problem for the fermionic Goldstone particles with spin 1/2. A successful solution of this problem led Volkov together with Akulov to the discovery of the extended super-Poincaré group (1972).

After that Volkov stated the idea of gauging the super-Poincaré group, and suggested I try to realize it. I accepted the suggestion with great enthusiasm, and we started. Then my earlier experience concerning the formalisms for high spin fields and gauge theories was very suitable and useful. We decided to consider the spontaneously broken extended super-Poincaré gauge group in order to study the Higgs effect for the Goldstone particles with spin 1/2 which had been recently introduced. Under this investigation very interesting and important features were revealed.[1,2]

- First of all, the real gauge fields with spin 3/2 were introduced as graviton superpartners. This fact was very unusual for those times, because till then the gauge fields possessed only integer spins. The graviton superpartners with spin 3/2 were later called the gravitino fields.

- Secondly, the extended super-Poincaré gauge group gives a principle possibility for the nontrivial unification of gravity with the interactions based on the internal symmetries.

- Thirdly, the Higgs effect for the Goldstone particles with spin 1/2, later called the super-Higgs effect, essentially differs from the Higgs effect for the Goldstone particles with spin 0. It results in not only the gravitinos becoming massive, but also the space-time changes its own metric and topological properties: a nonzero cosmological constant appears.

We have written down our action as a sum of five invariant terms with arbitrary constants before them. These terms contain respectively: the Einstein–Cartan action for gravity, a kinetic term for the gravitino fields, a mass term of the gravitinos, a cosmological term and a term for the Yang–Mills fields with spin 1.

Thus, we see that the main notions and ingredients of supergravity were formulated in our papers.[1,2]

Soon after and independently Wess and Zumino (1974) introduced a four-dimensional superconformal group generalizing the group of two-dimensional supersymmetric transformations present in the dual models. Then they proposed a supersymmetric model (1974) which appeared to be renormalizable.

The main results of our works were used afterwards as starting points for the development of supergravity in the works by Ferrara, Freedman, van Nieuwenhuizen[3] and Deser and Zumino.[4] In particular, their transformation rules of the vierbein, used for the graviton description, and gravitinos, coincided with ours. Explicit use of nonzero torsion, found in supersymmetric theories by us[5] and Zumino,[6] also was essential to simplify the derivation of supergravity in Ref. 4. Moreover, as Volkov explained,[7] pure N = 1 supergravity[3,4] can be deduced from our theory[1,2] with a special choice of the constants of our invariant terms.

The notions of superspace, introduced by Akulov and Volkov (1972), and superfields, proposed by Salam and Strathdee (1974), were used by Arnowitt, Nath and Zumino in 1975 to initiate a superfield approach to supergravity. In order to overcome the drawbacks they found at their first step in this direction, we[5] undertook, simultaneously with Zumino,[6] the development of the superspace approach to supergravity. By generalizing to superspace Cartan's methods in differential geometry, we revealed a nonzero torsion of superspace in the supersymmetric theories and found that the true homogeneous holonomy group for the superspace in supergravity is the Lorentz group. These two points are part of the basis of any version of supergravity.

Let me also mention our construction in the superfield formulation of a version of N = 1 supergravity in the linear approximation with the so-called new minimal set of auxiliary fields,[8] which we had found earlier than the old minimal set that was obtained by Ferrara, van Nieuwenhuizen, Stelle and West (1978).

So, I recalled the steps in supergravity which had been performed in our works in 1973–1977.

As the complete list of papers on the theme of my reminiscence is very huge and essentially exceeds this note, I refer only to our works and to those very closely related to them. A very extensive list of references concerning the development of supergravity can be found in Ref. 9. This list contains also our papers,[1,2,5,8] however, the references to the latter are omitted in the text of this review.

To finish my note, I want to cite a very surprising (for me) appraisal of supersymmetry given by Yuri A. Gol'fand during the Conference "Supersymmetry–85" at Kharkov State University in 1985. He said that supersymmetry did not justify his hopes to find a generalization of the Poincaré group such that its representations include particles of different masses. So, Gol'fand and Likhtman had missed their aim, but had instead found supersymmetry, every representation of which contains fields of different spins. Maybe, somebody will succeed in their original aim too, or maybe both aims at once. Time will tell.

In conclusion I would like to thank the organizers of this Symposium for the opportunity to present my recollections concerning the creation of the main ideas of supergravity at its early stages, which are usually not illuminated.

References

1. D.V. Volkov and V.A. Soroka, *JETP Lett.* **18** (1973) 312.
2. D.V. Volkov and V.A. Soroka, *Theor. Math. Phys.* **20** (1974) 829.
3. D.Z. Freedman, P. van Nieuwenhuizen and S. Ferrara, *Phys. Rev.* **D13** (1976) 3214.

4. S. Deser and B. Zumino, *Phys. Lett.* **62B** (1976) 335.

5. V.P. Akulov, D.V. Volkov and V.A. Soroka, *JETP Lett.* **22** (1975) 187.

6. B. Zumino, in Proceedings of the Conference on Gauge Theories and Modern Field Theory, Northeastern University, 1975, p. 255; CERN preprint TH. 2120.

7. D.V. Volkov, in Proceedings of the International Conference on The History of Original Ideas and Basic Discoveries in Particle Physics, Erice, Italy, 29 July – 4 August 1994.

8. V.P. Akulov, D.V. Volkov and V.A. Soroka, *Theor. Math. Phys.* **31** (1977) 285.

9. P. van Nieuwenhuizen, *Phys. Rep.* **68** (1981) 189.

SUPERGRAVITY AND THE QUEST FOR A UNIFIED THEORY[a]

SERGIO FERRARA

Theory Division, CERN,

CH-1211 Genéve 23, Switzerland

A recollection of some theoretical developments that preceded and followed the first formulation of supergravity theory is presented. Special emphasis is placed on the impact of supergravity on the search for a unified theory of fundamental interactions.

It is a great honor and pleasure to be invited to give this Dirac Lecture on the occasion of the 1994 Spring School on String, Gauge Theory and Quantum Gravity. In fact, this School is a continuation of a very successful series initiated by Prof. A. Salam in 1981. Together with J.G. Taylor and P. van Nieuwenhuizen I had the privilege of organizing the first two in the spring of 1981 and the fall of 1982.[1]

At that time, supergravity was in the mainstream of research, namely:

(i) Quantum properties of extended supergravities and their geometric structure;

(ii) Kaluza–Klein supergravity;

(iii) Models for particle physics phenomenology.

These topics were widely covered during the first two schools and workshops. Before going on to discuss supergravity and its subsequent development, let me briefly touch upon the steps taken in the preceding years, when supersymmetry in four dimensions was introduced.

Although the latter, with its algebraic structure, was first mentioned in 1971 by Gol'fand and Likhtman[2] and in early 1973 by Volkov and Akulov[3] (to explain the masslessness of the neutrino as a Goldstone fermion), it was really brought to the attention of theoretical particle physicists in the second half of 1973, by Wess and Zumino,[4] they had been inspired by a similar structure, found by

---

[a]The 1994 Dirac Prize Lecture, delivered at the International Centre for Theoretical Physics, Trieste, on 19 April, 1994.

Gervais and Sakita (1971),[5] already present in two dimensions, in the dual-spinor models constructed in 1971 by Neveu and Schwarz[6] and by Ramond.[7] The relevance of supersymmetry for quantum field theory, especially in view of its remarkable ultraviolet properties and its marriage with Yang–Mills gauge invariance, was soon established in early 1974.

It is nevertheless curious that it was only, at the time, rather isolated groups that delved into the subject, mainly in Europe: at CERN, the ICTP (Trieste), Karlsruhe, the ENS–Paris, Imperial College–London, Turin University, and essentially two in the United States: Caltech and Stony Brook. The same applies to supergravity and its ramifications in the early years, after its foundation in 1976.

Soon after the very first paper of Wess and Zumino,[8] a remarkable sequence of events occurred during 1974:

(i) The superspace formulation of supersymmetric field theories (Salam and Strathdee,[9] Wess, Zumino, and Ferrara[10]);

(ii) The discovery of non-renormalization theorems (Wess and Zumino,[11] Iliopoulos and Zumino,[12] Ferrara, Iliopoulos, and Zumino[13]).

(iii) The construction of supersymmetric Yang–Mills theories (Wess and Zumino[14] for the Abelian case; Ferrara and Zumino,[15] and Salam and Strathdee,[16] for the non-Abelian case);

(iv) The first construction of a renormalizable field theory model with spontaneously broken supersymmetry (Fayet and Iliopoulos[17]);

(v) The construction of a multiplet of currents, including the supercurrent and the stress energy tensor (Ferrara and Zumino[18]), which act as a source for the supergravity gauge fields and had an impact also later, in the classification of anomalies and in the covariant construction of superstring Lagrangians.

In the same year, quite independently of supersymmetry, Scherk and John Schwarz[19] proposed string theories as fundamental theories for quantum gravity and other gauge forces rather than for hadrons, turning the Regge slope from $\alpha' \sim \text{GeV}^{-2}$ to $\alpha' \sim 10^{-34} \text{ GeV}^{-2}$, the evidence being that any such theory contained a massless spin 2 state with interactions for small momenta as predicted by Einstein's theory of general relativity.

In the following year, many models with spontaneously broken supersymmetry and gauge symmetry were constructed, mainly by Fayet[20] and O'Raifeartaigh,[21] and $N = 2$ gauge theories coupled to matter, which were formulated by Fayet.[22]

This was a prelude to two important events, which took place just before and soon after the proposal of supergravity: the discovery of extended superconformal algebras (Ademollo *et al.*, November 1975[23]) and the finding of evidence for space-time supersymmetry in superstring theory (Gliozzi, Olive, and Scherk, GOS for short, September 1976 and January 1977[24]). In retrospect, these episodes had a great impact on the subsequent development of string theories in the mid-1980's.

The paper of Ademollo *et al.*, just a few months before supergravity was formulated, was inspired by the fact that it was possible, in higher dimensions ($D = 4$), by undoing the superspace coordinate θ_i with a counting index ($i = 1, ..., N$), to construct extended supersymmetries; indeed, a remarkable theory with $N = 2$ ($D = 4$) supersymmetry then had just been discovered by Fayet (September of 1985[22]). In $D = 4$, extended superconformal algebras were accompanied by U(N) gauge algebras [SU(4) for $N = 4$]. In $D = 2$, superconformal algebras are infinite-dimensional and $N = 2$ and $N = 4$ turned out to be accompanied by U(1) and SU(2) Kač–Moody gauge algebras. These algebras, at the time thought of as gauge-fixing of superdiffeomorphisms, were introduced to study new string theories with different critical dimensions.[25] In retrospect, this construction had a major impact on the classification of "internal" superconformal field theories, especially $N = 2$, as the quantum version of Calabi–Yau manifolds, and on its relation[26] with the existence of space-time supersymmetry in $D < D_{\text{crit}}$.

Meanwhile, in the spring of 1976,[27] supergravity was formulated by Freedman, van Nieuwenhuizen and the author, working at Ecole Normale and at Stony Brook. Soon after, a simplified version (first-order formulation) was presented by Deser and Zumino.[28]

While in the second formalism, the spin-3/2 four-fermion interaction has the meaning of a contact term (similar to seagull terms in scalar electrodynamics or non-Abelian gauge theories) required by fermionic gauge invariance, in the first-order formalism it has the meaning of a torsion contribution to the spin connection from "spin-3/2 matter." This discovery also implied that any supersymmetric system coupled to gravity should manifest local supersymmetry.

This observation eventually led some physicists to go deeper in string theory in order to explore whether the "dual spinor model" could accommodate target-space supersymmetry. The GOS paper (September 1976 and January 1977[24]) gave dramatic evidence for space-time supersymmetry in the superstring theory (called at that time the dual-spinor model)[29] by cutting out the G-odd parity states in the Neveu-Schwarz sector and comparing its bosonic spectrum with the fermion spectrum of the Ramond sector. In the proof, they used an identity that had been proved by Jacobi in 1829 *(Aequatio identica satis abstrusa)*! This paper came out following some sequential developments in supergravity, just after its first construction in the spring of 1976, namely the first matter-coupling to Maxwell theory (Ferrara, Scherk, and van Nieuwenhuizen, August 1976[30]) to Yang–Mills theory and chiral multiplets[31] and the first formulation of extended supergravity [$N = 2$] (Ferrara and van Nieuwenhuizen, September 1976[32]). It is interesting to note that two of the GOS authors (G and S) also took part in some of the above-mentioned supergravity papers.

The hypothesis of GOS (later proved in great detail by Green and John Schwarz[33]) also implied the existence of an $N = 4$ Yang–Mills theory, eventually coupled to an $N = 4$ extended supergravity. This was implied by a dimensional reduction of the ten-dimensional spectrum. The full $N = 4$ supergravity contained in this reduction was found a year later (Cremmer, Scherk, and Ferrara, December 1977[34]) and it was shown to contain an SU(4) × SU(1,1) symmetry. Meanwhile, three other important developments were announced at the end of 1976. The construction of $N = 3$ supergravity (Freedman; Ferrara, Scherk, and Zumino, November 1976[35]) and the discovery of (Abelian and non-Abelian) duality symmetries, generalizing the electromagnetic duality $F \rightarrow \tilde{F}$ in $N = 2$ [U(1)] and $N = 3$ [U(3)] supergravity (December 1976[36]). This duality generalizes to SO(6) × SU(1,1) in pure $N = 4$ supergravity and to SO(6,n)× SU(1,1) in $N = 4$ supergravity coupled to n matter (Yang–Mills) multiplets.

In retrospect, these symmetries play a crucial role in compactified superstrings, where the manifold

$$\frac{SO(6, N)}{SO(6) \times SO(N)} \times \frac{SU(1, 1)}{U(1)}$$

(modded out by some discrete symmetries) describes the moduli space of toroidally compactified ten-dimensional strings, according to the analysis of Narain, Sarmadi, and Witten[37]).

In September 1976, also the covariant world-sheet formulation of the spinning string was presented in two papers[38] by Brink, Di Vecchia and Howe and by Deser and Zumino. In retrospect this can be considered as a crucial ingredient for the Polyakov formulation[39] of spinning strings with arbitrary world-sheet topology. In this respect, $(p + 1)$ supergravity is necessary for the consistent formulation of any p-dimensional extended object coupled to fermions.

In the subsequent years all higher extended four-dimensional supergravities with $N = 5$, 6 and 8 were constructed.

The maximally extended supergravity ($N = 8$) was found by Cremmer and Julia,[40] by dimensional reduction of eleven-dimensional supergravity previously obtained by the same authors with Scherk (1978[41]), and its gauged version, accompanied with an SO(8) Yang–Mills symmetry, by de Wit and Nicolai (1982[42]). Gell-Mann had earlier observed that SO(8) cannot accommodate the observable gauge symmetry SU(3) × SU(2) × U(1) of electroweak and strong interactions. However, it was later observed by Ellis, Gaillard, Maiani and Zumino (1982[43]) that a hidden local SU(8) symmetry (found by Cremmer and Julia) could be identified as a viable Grand Unified Theory (GUT) for non-gravitational interactions. The basic assumption was that the degrees of freedom of the SU(8) gauge bosons could be generated at the quantum level, as it was known to occur in certain two-dimensional σ-models, following the analysis of Di Vecchia, D'Adda and Lüscher.[44] However, contrary to two-dimensional σ-models, which are renormalizable and therefore consistent quantum field theories, it turned out later that $N = 8$ supergravity in $D = 4$, which is also a kind of generalized σ-model, is unlikely to enjoy a similar property. This is one of the reasons why this attempt was abandoned. Another reason was closely related to the forthcoming string revolution, when Green and Schwarz (GS) (1984[45]) proved that $D = 10$, $N = 1$ supergravity, coupled to supersymmetric Yang–Mills matter, could be embedded in a consistent superstring theory for a particular choice of gauge groups (SO(32) and $E_8 \times E_8$).

The GOS and GS papers gave strong evidence that superstrings consistent with space-time supersymmetry containing supergravity + matter (rather than pure higher extended supergravity), in the massless sector, were a possible candidate for a theory of quantum gravity, encompassing the other gauge interactions and free from unphysical ultraviolet divergences. On the contrary, in the context of point-field theories, these systems, even if the symmetries dictated in an almost unique way all the couplings, were found to be non-renormalizable, already at one loop, when standard perturbative techniques were applied to them (Grisaru, van Nieuwenhuizen, and Vermaseren, 1976[46]). Indeed it was later shown that this was also the case for pure supergravities at and beyond three loops. [These theories had, however, the remarkable property of being one- and two-loop finite (Grisaru, van Nieuwenhuizen, and Vermaseren,[46] Grisaru,[47] Tomboulis[47]).] Pioneering work, in the late 70's, was also the analysis of spontaneous supersymmetry breaking in global and local supersymmetry. In rigid supersymmetry, Fayet[48] opened the way to the construction of the minimal supersymmetric extension of the Standard Model (MSSM), which in particular demanded two Higgs doublets. However, the gauge and supersymmetry breaking introduced by him required more degrees of freedom than the MSSM.

When supersymmetry is gauged, i.e. in supergravity, the supersymmetric version of the Higgs mechanism appears (super-Higgs), i.e. the goldstino is eaten up by the spin-3/2 gravitino (the gauge fermion of supergravity, the partner of the gravitons), which then becomes massive.

The possibility of having spontaneously broken supergravity with vanishing cosmological constant was shown by Deser and Zumino (April 1977[49]) and proved in detail by Cremmer *et al.* (August 1978[50]), by studying the most general matter coupling to $N = 1$ supergravity for a chiral multiplet, whose superpotential triggers a non-vanishing gravitino mass. The Higgs effect for Goldstone fermions had also been considered earlier by Volkov and Soroka.[51]

Another important result at that time, found by Zumino (August 1979[52]), was the fact that the most general supergravity couplings of chiral multiplets (with two-derivative action) were described by Kählerian σ-models.

Again, in retrospect, this Kählerian structure and the generalization thereof have played a role in superstring theory from both the world-sheet and target-space points of view.

Although in the 1970's the work done in supersymmetric models for particle physics (using renormalizable Lagrangians with spontaneously broken supersymmetry) and that towards a deeper understanding of the structure of supergravity theories (off-shell formulations, matter couplings, etc.) went in parallel, with small intersections, they came closer and became eventually deeply connected after two major developments were made in the early 1980's.

The first was the call made upon supersymmetry breaking near the electroweak scale, to solve the so-called hierarchy problem of gauge theories with fundamental Higgs scalars (Gildener, Weinberg; Veltman; Witten; Maiani).[53,54]

This development and general properties of criteria for supersymmetry breaking, contained in two pivotal papers by Witten (April 1981, January 1982[54]), opened up the field of supersymmetry and supergravity as main stream research in the United States and in the rest of the world.

The hierarchy problem is connected to the unnaturalness of the hierarchy E_F/E_X (E_F being the Fermi scale) in any renormalizable theory with fundamental scalars, whose vacuum expectation value triggers the electroweak gauge symmetry breaking at a scale E_F much lower than any other (cut-off) scale E_X.

This is due to the quadratic dependence on the cut-off Λ of the effective potential, which, at one loop, manifests itself in a term

$$\sum_{J_i}(-)^{2J_i}(2J_i+1)\mathcal{M}_{J_i}^2(\phi)\Lambda^2 \ ,$$

where $M_{J_i}^2(\phi)$ are the (scalar) field-dependent masses of particle species with spin J_i. In an arbitrary supersymmetric renormalizable field theory with no traceful Abelian gauge group factor, the expression multiplying Λ^2 identically vanishes (owing to the special relation between boson and fermion couplings), as was shown by Girardello, Palumbo and the author (April 1979[55]). This is also true for matter-coupled $N = 1$ supergravity with a single chiral multiplet on a flat Kähler manifold (1978[50]) and in spontaneously broken extended supergravity via the Scherk–Schwarz mechanism (1979[56]).

However, a closer look at boson–fermion mass matrices revealed that this property made models previously considered by Fayet more problematic, since they tended either to give an unrealistic spectrum with some scalar superpartners of quarks and leptons lighter than their fermionic counterparts, or to need a traceful additional U(1) gauge interaction, plagued with triangular anomalies. Cancelling these anomalies usually needed extra fields, which eventually allowed vacua with broken color or charge symmetry.

However, when the most general coupling of $N = 1$ supergravity to an arbitrary matter system, with arbitrary gauge interactions, became available (Cremmer, Ferrara, Girardello, and Van Proeyen, 1982[57]), it was realized that, provided $m_{3/2} \ll M_{\rm Pl}$ and possibly $\simeq O({\rm TeV})$, mass terms for any observable scalar $O(m_{3/2})$ were easily generated, thus resolving the partner–spartner splitting problem, which generally occurred in spontaneously broken rigid theories.

There is an alternative way of phrasing this: in the Fayet-type models, the goldstino has coupling to the observable sector $O(1)$ and the gravitino mass is very tiny, $m_{3/2} \sim 10^{-13}$ GeV, while in supergravity models with $m_{3/2} \gtrsim O(m_{\rm Z})$ the goldstino coupling is highly suppressed $[O(m_{3/2}/M_{\rm Pl})]$, which implies that the gravitino only carries gravitational interactions (Fayet[58]).

In the limit in which $m_{3/2}$ is kept fixed and couplings $O(1/M_{\rm Pl})$ are neglected, spontaneously broken supergravity models behave as globally supersymmetric models with softly broken terms, i.e. terms with dimension ≤ 3, which do not induce quadratic divergences in the low-energy effective theory.

These terms had been classified in 1981 by Girardello and Grisaru.[59] A generalization of non-renormalization theorems for superpotential terms in a generic theory were also found using superspace techniques, by Grisaru, Siegel, and Roček (June 1979[60]).

Softly broken terms and renormalization theorems were used to construct viable supersymmetric GUTs, including the MSSM as their low-energy effective theory, with no hierarchy problem (the first of these was constructed by Georgi and Dimopoulos in the summer of 1981[61]). Soon after, realistic electroweak and GUT models, with spontaneously broken supersymmetry triggered by the supergravity couplings at the tree level, were constructed (Barbieri, Ferrara, and Savoy, 1982,[62] Chamseddine, Nath, and Arnowitt, 1982,[63] Hall, Lykken, and Weinberg, 1983[64]). A general feature of these models is that the messengers of supersymmetry breaking to the observable sector (encompassing electroweak and strong interactions) are a set of neutral chiral multiplets (called the hidden sector), which have only gravitational interactions and decouple from the low-energy theory; in the latter, the only trace of them is to produce the soft-breaking terms, then having the effect of modifying the supertrace formula of global supersymmetry with an additional (field-independent) constant (with no physical consequences on the theory decoupled from gravity).

Nowadays, in the MSSM, the electroweak symmetry is broken through radiative corrections, through a Coleman–Weinberg mechanism, while supersymmetry is broken at the tree level through the soft-breaking terms.

Considering the initial condition for the couplings as given at the Planck scale and evolving them through the renormalization group equations (Ibàñez, Ross in 1981,[65] Alvarez-Gaumé, Polchinski, and Wise in 1982[66]) in a region of the parameter space, the electroweak symmetry was indeed found to be spontaneously broken with a Higgs mass of the same order of magnitude as the gravitino mass. There is a particular subclass of spontaneously broken supergravity models, called no-scale supergravities (Cremmer, Ferrara, Kounnas and Nanopoulos, 1983,[67] Ellis *et al.*,[68]), where the supergravity-breaking scale $m_{3/2}$ is arbitrary at the tree level (owing to flat directions in the superpotential). In these models, radiative corrections may generate the hierarchy $m_{3/2} = M_{\rm Pl} \ e^{-c/g^2}$, then explaining how a small scale can be created in a theory whose only original dimensionful scale is $M_{\rm Pl}$.

It was later shown by Witten[69] that many four-dimensional superstring models have, in the point-field limit, a no-scale structure; therefore, after supersymmetry breaking, they may give rise to a dynamical hierarchy.

Nowadays almost every particle physicist knows what $\tan\beta, A, B$ represent in the general parametrization of the soft-breaking terms of the MSSM.

The second breakthrough was on physics at the Planck scale (Green, Schwarz, September 1984,[45]) namely the discovery of anomaly-free ten-dimensional supergravity coupled to Yang–Mills matter or consistent superstring theories, for specific gauge group choices (in open and heterotic strings) (Gross, Harvey, Martinec, and Rohm, November 1984[70]). Heterotic string theories, after suitable compactification of six extra dimensions (on some compact manifolds with special properties), led to $N = 1$ effective supergravity theories, with a spectrum of charged chiral multiplets (chiral with respect to the surviving gauge group $G' \supset SU(3) \times SU(2) \times U(1)$ (after compactification) and accommodating families of quarks and leptons with the $SU(3) \times SU(2) \times U(1)$ assignment of the Standard Model.

The use of ten-dimensional Yang–Mills fields, prior to compactification, is crucial to overcome previous difficulties encountered in Kaluza–Klein supergravities (Freund and Rubin, 1980,[71] Witten, 1981,[72] Duff, Nilsson, and Pope, 1986[73]), where attempts were made at obtaining the $SU(3) \times SU(2) \times U(1)$ gauge symmetries from the isometries of the internal manifold. In fact, even if in some cases the desired gauge group could be obtained (Witten, 1981,[72] Castellani, D'Auria, and Fré, 1984[74]), these failed because the resulting fermion spectrum was not chiral with respect to the electroweak gauge symmetry.

In models where supersymmetry breaking occurs via a non-trivial dilaton superpotential, the neutral fields coming from the internal components of the metric tensor G_{IJ} are natural candidates for flat directions, at least in the limit of manifolds that are "large" with respect to the string scale.

In particular, in four-dimensional heterotic superstring theories, compactified on Calabi–Yau manifolds (Candelas, Horowitz, Strominger, and Witten, January 1985[75]), or their "orbifold limit,"[76] a natural identification of the hidden sector occurs with a set of "moduli fields", which parametrize the deformations of the Kähler class and complex structure of the manifold (generalization of radial deformations of simple tori).[77] A popular scenario for a non-perturbative dilaton superpotential is the gaugino-condensation mechanism (Ferrara, Girardello, and Nilles[78]) in the hidden sector gauge group (Derendinger, Ibáñez, and Nilles; Dine, Rohm, Seiberg, and Witten[79]). The fact that some moduli remain large (with respect to the string scale) could result in a sliding gravitino mass, which could then be stabilized through radiative corrections in the observable sector with a dynamical suppression with respect to $M_{\rm Pl}$.

In recent years a suggestion has been made (Duff,[80] Strominger[81]) that strings may, in the strong coupling regime, have a simpler formulation in terms of a dual theory (five-brane) in the weak coupling. These theories, in $D = 10$, have the same field theory limit, namely, $D = 10$ supergravity (which is unique because of supersymmetry), but electrically charged massive string states correspond to "magnetically" charged five-brane states (and vice versa) with a similar relation as it occurs between electric and magnetic charge in Dirac monopole quantization.[82]

This is an explicit manifestation of the general fact that a $(p + 1)$ form gauge field is, in D dimensions, naturally coupled to a p-dimensional extended object, and that its "dual" potential (which is a $D - p - 3$ form) is naturally coupled to a $D - p - 4$ extended object. From topological arguments, similar to Dirac's, the product of the two couplings must be quantized.

In toroidal compactifications, it has indeed been shown (Sen and Schwarz[83]) that the spectrum of both electrically and magnetically charged states (the latter obtained from the low-energy effective field theory, i.e. an $N = 4$ supergravity coupled to Yang–Mills) satisfies an SL(2,Z) duality for the dilaton chiral multiplet $S = (1/g^2) + i\theta$ (here g^2 is the field-dependent gauge coupling and θ is the field-dependent "θ-angle"). This may therefore suggest a "modular-invariant" dilaton potential.[84] This symmetry is similar to the "moduli duality," which occurs in weakly coupled strings as a consequence of the world-sheet non-trivial topology.[85]

This approach is worth exploring, even if it is difficult to imagine that a microscopic, consistent quantum theory describing five-brane propagation could possibly exist.

Finally, let me conclude by making some remarks about the possible indirect experimental signals, indicating that supersymmetry may be just around the corner. With an optimistic attitude, these are:

1) The non-observation of proton decay within the limit of a lifetime of 10^{32} years in the main channel $p \to \pi^0 e^+$, excluding minimal GUT unification.

2) The LEP precision measurements, which are incompatible with gauge-coupling unification for conventional minimal GUTs, but are in reasonable agreement with minimal supersymmetric GUTs, with supersymmetry broken at the TeV scale.

3) The top Yukawa coupling, unusually large with respect to the one of other quarks and leptons.

4) The possible resolution of the dark-matter problem, with some of the neutral supersymmetric particles as natural dark-matter candidates.

Although none of these facts is *per se* a compelling reason for supersymmetry and may find alternative explanations, it is fair to say that they can all be explained in the context of a supersymmetric extension of ordinary gauge theories.

Whatever the final theory (strings?) for quantum gravity will be, supergravity[86] remains a deep and non-trivial extension of the principle of general covariance and Yang–Mills gauge symmetry, which played such an important role in the description of natural phenomena.

Let us hope that nature has used this fascinating structure!

References

1. S. Ferrara and J.G. Taylor (eds.), *Supergravity 81* (Cambridge Univ. Press, Cambridge, 1981);
S. Ferrara, J.G. Taylor and P. van Nieuwenhuizen (eds.), *Supersymmetry and Supergravity 82* (World Scientific, Singapore, 1982).

2. Yu.A. Gol'fand and E.P. Likhtman, *JETP Lett.* **13** (1971) 323.

3. D.V. Volkov and V.P. Akulov, *Phys. Lett.* **46B** (1973) 109.

4. B. Zumino, in *Renormalization and invariance in quantum field theory*, ed. E. Caianello (Plenum Press, New York, 1974), p. 367.

5. J.-L. Gervais and B. Sakita, *Nucl. Phys.* **B34** (1971) 633.

6. A. Neveu and J. Schwarz, *Nucl. Phys.* **B31** (1971) 86.

7. P. Ramond, *Phys. Rev.* **D3** (1971) 2415.

8. J. Wess and B. Zumino, *Nucl. Phys.* **B70** (1974) 39.

9. A. Salam and J. Strathdee, *Nucl. Phys.* **B76** (1974) 477.

10. S. Ferrara, B. Zumino and J. Wess, *Phys. Lett.* **51B** (1974) 239.

11. J. Wess and B. Zumino, *Phys. Lett.* **49B** (1974) 52.

12. J. Iliopoulos and B. Zumino, *Nucl. Phys.* **B76** (1974) 310.

13. S. Ferrara, J. Iliopoulos and B. Zumino, *Nucl. Phys.* **B77** (1974) 413.

14. J. Wess and B. Zumino, *Nucl. Phys.* **B78** (1974) 1.

15. S. Ferrara and B. Zumino, *Nucl. Phys.* **B79** (1974) 413.

16. A. Salam and J. Strathdee, *Phys. Lett.* **51B** (1974) 353.

17. P. Fayet and J. Iliopoulos, *Phys. Lett.* **51B** (1974) 461.

18. S. Ferrara and B. Zumino, *Nucl. Phys.* **B87** (1975) 207.

19. J. Scherk and J. Schwarz, *Nucl. Phys.* **B81** (1974) 118.

20. P. Fayet, *Phys. Lett.* **58B** (1975) 67.

21. L. O'Raifeartaigh, *Nucl. Phys.* **B6** (1975) 331.

22. P. Fayet, *Nucl. Phys.* **B113** (1976) 135.

23. M. Ademollo, L. Brink, A. D'Adda, R. D'Auria, E. Napolitano, S. Sciuto, E. del Giudice, P. Di Vecchia, S. Ferrara, F. Gliozzi, R. Musto and R. Pettorino, *Phys. Lett.* **62B** (1976) 105.

24. G. Gliozzi, J. Scherk and D. Olive, *Phys. Lett.* **65B** (1976) 282; *Nucl. Phys.* **B122** (1977) 253.

25. M. Ademollo, L. Brink, A. D'Adda, R. D'Auria, E. Napolitano, S. Sciuto, E. Del Giudice, P. Di Vecchia, S. Ferrara, F. Gliozzi, R. Musto, R. Pettorino and J. Schwarz, *Nucl. Phys.* **B111** (1976) 77.

26. D. Gepner, *Nucl. Phys.* **B296** (1988) 57; *Phys. Lett.* **199B** (1987) 477;
W. Lerche, D. Lüst and A.N. Schellekens, *Nucl. Phys.* **B287**

(1987) 477;

H. Kawai, D.C. Lwellen and S.-H.H. Tye, *Nucl. Phys.* **B298** (1987) 1;

L. Antoniadis, C.P. Bachas and C. Kounnas, *Nucl. Phys.* **B289** (1987) 87.

27. D.Z. Freedman, P. van Nieuwenhuizen and S. Ferrara, *Phys. Rev.* **D13** (1976) 3214.

28. S. Deser and B. Zumino, *Phys. Lett.* **62B** (1976) 335.

29. This nomenclature was inherited from the "dual models" that had started with the seminal paper by
G. Veneziano, *Nuovo Cim.* **57A** (1968) 190.

30. S. Ferrara, J. Scherk and P. van Nieuwenhuizen, *Phys. Rev. Lett.* **37** (1976) 1035.

31. S. Ferrara, F. Gliozzi, J. Scherk and P. van Nieuwenhuizen, *Nucl. Phys.* **B117** (1976) 333;
D.Z. Freedman and J.H. Schwarz, *Phys. Rev.* **D15** (1977) 1007;
S. Ferrara, D.Z. Freedman, P. van Nieuwenhuizen, P. Breitenlohner, F. Gliozzi and J. Scherk, *Phys. Rev.* **D15** (1973) 1013.

32. S. Ferrara and P. van Nieuwenhuizen, *Phys. Rev. Lett.* **37** (1976) 1669.

33. M.B. Green and J.H. Schwarz, *Nucl. Phys.* **B181** (1981) 502;
Phys. Lett. **109B** (1982) 444 and **136B** (1984) 367.

34. E. Cremmer, J. Scherk and S. Ferrara, *Phys. Lett.* **68B** (1977) 234.

35. D.Z. Freedman, *Phys. Rev. Lett.* **38** (1977) 105;
S. Ferrara, J. Scherk and B. Zumino, *Phys. Lett.* **66B** (1977) 35.

36. S. Ferrara, J. Scherk and B. Zumino, *Nucl. Phys.* **B121** (1977) 393.

37. K. Narain, *Phys. Lett.* **169B** (1986) 41;
K. Narain, M.H. Sarmadi and E. Witten, *Nucl. Phys.* **B279** (1986) 96.

38. L. Brink, P. Di Vecchia and P. Howe, *Phys. Lett.* **65B** (1976) 471;
S. Deser and B. Zumino, *Phys. Lett.* **65B** (1976) 369.

39. A.M. Polyakov, *Phys. Lett.* **103B** (1981) 207 and 211.
40. E. Cremmer and B. Julia, *Phys. Lett.* **80B** (1978) 48; *Nucl. Phys.* **B159** (1979) 141.
41. E. Cremmer, B. Julia and J. Scherk, *Phys. Lett.* **76B** (1978) 409.
42. B. de Wit and H. Nicolai, *Nucl. Phys.* **B208** (1982) 323.
43. J. Ellis, M.K. Gaillard, L. Maiani and B. Zumino, in *Unification of fundamental interactions*, eds. S. Ferrara, J. Ellis and P. van Nieuwenhuizen [Plenum Press (Ettore Majorana Series), New York, 1980], p. 69;
    J. Ellis, M.K. Gaillard and B. Zumino, *Phys. Lett.* **94B** (1980) 343.
44. A. D'Adda, P. Di Vecchia and M. Lüscher, *Nucl. Phys.* **146** (1978) 63 and **B152** (1979) 125.
45. M. Green and J. Schwarz, *Phys. Lett.* **149B** (1984) 117.
46. M.T. Grisaru, P. van Nieuwenhuizen and J.A.M. Vermaseren, *Phys. Rev. Lett.* **37** (1976) 1662;
    P. van Nieuwenhuizen and J.A.M. Vermaseren, *Phys. Lett.* **65B** (1976) 263.
47. M.T. Grisaru, *Phys. Lett.* **B66** (1977) 75;
    E.T. Tomboulis, *Phys. Lett.* **67B** (1977) 417.
48. P. Fayet, *Phys. Lett.* **69B** (1977) 489.
49. S. Deser and B. Zumino, *Phys. Rev. Lett.* **38** (1977) 1433.
50. E. Cremmer, B. Julia, J. Scherk, S. Ferrara, L. Girardello and P. van Nieuwenhuizen, *Phys. Lett.* **79B** (1978) 231; *Nucl. Phys.* **B147** (1979) 105.
51. D.V. Volkov and V.A. Soroka, *JETP Lett.* **18** (1973) 312.
52. B. Zumino, *Phys. Lett.* **87B** (1979) 203.
53. E. Gildener and S. Weinberg, *Phys. Rev.* **D13** (1976) 3333;
    S. Weinberg, *Phys. Lett.* **82B** (1979) 387;
    M. Veltman, *Acta Phys. Polon.* **B12** (1981) 437;
    L. Maiani, *Proc. Summer School of Gif-sur-Yvette* 1979 (IN2P3, Paris, 1979), p. 1.
54. E. Witten, *Nucl. Phys.* **B185** (1981) 513 and **B202** (1982) 253.
55. S. Ferrara, L. Girardello and F. Palumbo, *Phys. Rev.* **D20** (1979) 403.

56. J. Scherk and J. Schwarz, *Nucl. Phys.* **B153** (1979) 61;
    E. Cremmer, J. Scherk and J. Schwarz, *Phys. Lett.* **82B** (1979) 60.
57. E. Cremmer, S. Ferrara, L. Girardello and A. van Proeyen, *Phys. Lett.* **116B** (1982) 231; *Nucl. Phys.* **B212** (1983) 413.
58. P. Fayet, *Phys. Lett.* **70B** (1977) 461.
59. L. Girardello and M.T. Grisaru, *Nucl. Phys.* **B194** (1982) 65.
60. M.T. Grisaru, W. Siegel and M. Roček, *Nucl. Phys.* **B159** (1979) 429.
61. S. Dimopoulos and H. Georgi, *Nucl. Phys.* **B193** (1981) 150.
62. R. Barbieri, S. Ferrara and C. Savoy, *Phys. Lett.* **199B** (1982) 343.
63. A.H. Chamseddine, R. Arnowitt and P. Nath, *Phys. Rev. Lett.* **49** (1982) 970.
64. L. Hall, J. Lykken and S. Weinberg, *Phys. Rev.* **D27** (1983) 2359.
65. L. Ibáñez and G.G. Ross, *Phys. Lett.* **105B** (1981) 439 and **110B** (1982) 215.
66. L. Alvarez-Gaumé, J. Polchinski and M. Wise, *Nucl. Phys.* **B221** (1983) 495.
67. E. Cremmer, S. Ferrara, C. Kounnas and D.V. Nanopoulos, *Phys. Lett.* **133B** (1983) 61.
68. J. Ellis, A.B. Lahanas, D.V. Nanopoulos and K. Tamvakis, *Phys. Lett.* **134B** (1984) 429;
    J. Ellis, C. Kounnas and D.V. Nanopoulos, *Nucl. Phys.* **B247** (1984) 373.
69. E. Witten, *Phys. Lett.* **155B** (1985) 151.
70. D.J. Gross, J.A. Harvey, E. Martinec and R. Rohm, *Phys. Rev. Lett.* **54** (1985) 502.
71. P.G.O. Freund and M.A. Rubin, *Phys. Lett.* **97B** (1980) 233.
72. E. Witten, *Nucl. Phys.* **B186** (1981) 412.
73. M.J. Duff, B.E.W. Nilsson and C.N. Pope, *Phys. Rep.* **130** (1986) 1.
74. L. Castellani, R. D'Auria, and P. Fré, *Nucl. Phys.* **B239** (1984) 60.
75. P. Candelas, G. Horowitz, A. Strominger and E. Witten, *Nucl. Phys.* **B258** (1985) 46.

76. L. Dixon, J.A. Harvey, C. Vafa and E. Witten, *Nucl. Phys.* **B261** (1985) 678.

77. N. Seiberg, *Nucl. Phys.* **B303** (1988) 286;
    S. Cecotti, S. Ferrara and L. Girardello, *Int. J. Mod. Phys.* **4** (1989) 2475;
    P. Candelas, P. Green and T. Hübsch, *Nucl. Phys.* **B330** (1990) 49;
    P. Candelas and X.C. de la Ossa, *Nucl. Phys.* **B342** (1990) 246;
    S. Ferrara and A. Strominger, in *String '89*, eds. R. Arnowitt, R. Brijan, M.J. Duff and D.V. Nanopoulos (World Scientific, Singapore, 1989), p. 245;
    A. Strominger, *Comm. Math. Phys.* **133** (1990) 42;
    L. Dixon, V.S. Kaplunovsky and J. Louis, *Nucl. Phys.* **B329** (1990) 27;
    S. Cecotti and C. Vafa, *Nucl. Phys.* **B367** (1991) 359;
    A. Ceresole, R. D'Auria, S. Ferrara, W. Lerche and J. Louis, *Int. Mod. Phys.* **8A** (1993) 79.

78. S. Ferrara, L. Girardello and H.P. Nilles, *Phys. Lett.* **125B** (1983) 457.

79. J.-P. Derendinger, L.E. Ibáñez and H.P. Nilles, *Phys. Lett.* **155B** (1985) 65;
    M. Dine, R. Rohm, N. Seiberg and E. Witten, *Phys. Lett.* **156B** (1985) 55.

80. M.J. Duff, *Class. Quant. Grav.* **5** (1988) 189;
    M.J. Duff and J.X. Lu, *Phys. Rev. Lett.* **66** (1991) 1402.

81. A. Strominger, *Nucl. Phys.* **B343** (1990) 167;
    C.G. Callan, J.A. Harvey and A. Strominger, *Nucl. Phys.* **B359** (1991) 611.

82. C. Montonen and D. Olive, *Phys. Lett.* **B72** (1977) 117.

83. A. Sen, *Mod. Phys. Lett.* **A8** (1993) 2023;
    J.H. Schwarz and A. Sen, *Nucl. Phys.* **B411** (1994) 35; *Phys. Lett.* **B312** (1993) 105.

84. A. Font, L. Ibáñez, D. Lüst and F. Quevedo, *Phys. Lett.* **B249** (1990) 35.

85. K. Kikkawa and M. Yamasaki, *Phys. Lett.* **149B** (1984) 357;
    N. Sakai and L. Senda, *Progr. Theor. Phys.* **75** (1986) 692;

V.P. Nair, A. Shapere, A. Strominger and F. Wilczek, *Nucl. Phys.* **B287** (1987) 402;

A. Giveon, E. Rabinovici and G. Veneziano, *Nucl. Phys.* **B322** (1989) 167;

A. Shapere and F. Wilczek, *Nucl. Phys.* **B320** (1989) 167;

M. Dine, P. Huet and N. Seiberg, *Nucl. Phys.* **B322** (1989) 301;

J. Molera and B. Ovrut, *Phys. Rev.* **D40** (1989) 1150;

J. Lauer, J. Mas and H.P. Nilles, *Phys. Lett.* **B226** (1989) 251; *Nucl. Phys.* **B351** (1991) 353;

W. Lerche, D. Lüst and N.P. Warner, *Phys. Lett.* **B231** (1989) 417;

M. Duff, *Nucl. Phys.* **B335** (1990) 610;

A. Giveon and M. Porrati, *Phys. Lett.* **B246** (1990) 54; *Nucl. Phys.* **B355** (1991) 442;

A. Giveon, N. Malkin and E. Rabinovici, *Phys. Lett.* **B238** (1990) 57;

J. Erler, D. Jungnickel and H.P. Nilles, *Phys. Lett.* **B276** (1992) 303;

S. Ferrara, D. Lüst, A. Shapere and S. Theisen, *Phys. Lett.* **B233** (1989) 147;

L. Dixon, V. Kaplunovsky and J. Louis, *Nucl. Phys.* **B355** (1991) 649;

G.L. Cardoso and B. Ovrut, *Nucl. Phys.* **B369** (1992) 351 and **B392** (1993) 315;

J.-P. Derendinger, S. Ferrara, C. Kounnas and F. Zwirner, *Nucl. Phys.* **B372** (1992) 145;

L. Antoniadis, K. Narain and T. Taylor, *Phys. Lett.* **B267** (1991) 37;

L. Antoniadis, E. Gavai and K. Narain, *Nucl. Phys.* **B365** (1992) 93.

86. Reviews on the developments of supergravity in the 80's can be found in:
P. van Nieuwenhuizen, *Phys. Rep.* **68** (1981) 189;
H.P. Nilles, *Phys. Rep.* **110** (1984) 1;
S. Ferrara, Vol. 2 of Supersymmetry (North Holland, Amsterdam/World Scientific, Singapore, 1987).

4

The Pioneers

A GLIMPSE

JOHN STRATHDEE
Wellington, New Zealand

At this point all that comes to my mind is my annoyance at having our note[a] describing the unitary representations rejected by Physical Review Letters as being of "insufficient interest." I suppose I really suspected that the editor was quite right, and so was annoyed with myself for getting excited about such a far-fetched speculation.

Of course, I now realize, like everyone else, that the subject of supersymmetry is truly interesting ...

April 29, 2000

---

[a]The paper A. Salam and J. Strathdee, *Unitary Representations of Supergauge Symmetries*, was published in Nucl. Phys. **B80**, 499, 1974. The authors develop the ideas put forward in their previous paper (Nucl. Phys. **B76**, 477, 1974) where the superfield formalism had been introduced. They present an algebraic construction of the unitary representations of the super-Poincaré algebra. In Nucl. Phys. **B80**, 499, 1974 they also introduce what is called in modern terminology "extended superalgebras," and find some representations of the extended superalgebras. – Editors' note.

RECOLLECTIONS OF A YOUNG CONTRIBUTOR

MARTIN F. SOHNIUS

Novell European Services Ltd.

1, Arlington Square, Bracknell, England

Msohnius@novell.com

Having been asked to write some "personal recollections" of the early days of supersymmetry research, I must start with the peculiar apology that they are indeed going to be recollections! Literally. You see, being too young to be writing my memoirs quite yet, I do not have "my papers ordered" to do proper research into my own past. On the other hand, I have been "retired" from academic research for exactly 12 years and 13 days (by the date of the symposium[a]) – just long enough not to have names, institutions or, indeed, the latest gossip at my fingertips. As well as to have forgotten a good deal of the physics, and not to be quite sure whether what I do remember was ever correct, or is still considered to be correct. I am sitting here writing in an office of Novell Inc. with lovely high-bandwidth access to the Internet, but no access at all to a library containing Physics Letters or Nuclear Physics B. Sadly, 30-year old high-energy physics papers tend not to be on-line! Which, of course, means that my current "paper" will not have any references, and will mention names only where my middle-aged brain can recall both them and at least a fair approximation of their spelling. It'll all be quite personal.

In fact, I don't even have my own publication list handy. When Misha Shifman wrote in his invitation email that my paper in Nucl. Phys. **B88**, 257 (1975) would qualify me as a "pioneer of supersymmetry," I was at a loss: which of those early papers might he be referring to? I could, of course, guess (I didn't write that many famous papers, after all), but I wanted to make sure, and was therefore rather dismayed when the first thing "Yahoo!" came up with was a Japanese web page with the phrase "Nucl. Phys. **B88**, 257 (1975)" as the sole Latin content in what was otherwise a sea of Kanji. Well, in the end I could confirm, on the Web, the identity of that reference

[a]International Symposium "30 Years of Supersymmetry," Theoretical Physics Institute, University of Minnesota, October 13–15, 2000.

as the paper I wrote with Rudolf Haag and Jan Lopuszański about graded extensions of the Poincaré and conformal algebras (as, with hindsight, I would title it today). Actually, I always prefer to say "the paper that Haag and Lopuszański wrote with me" – it more accurately reflects the relationship of a young graduate student to a pair of world-renowned senior physicists. The following recollections of how it all came about are necessarily from a very personal point of view – the point of view, moreover, of a Ph.D. student with not much of an inkling of what was going on in the big world of physics outside Julius Wess' Karlsruhe Institute. So, in the end I have come full circle: I did not know then what was going on in physics, and here I am back at that same state of affairs today!

Which, of course, gives me the immense advantage of a fool's licence: none of you will in the near future referee any paper or grant application of mine, nor do I need to compete with any of you folks anymore. (I'll stick to competing with Bill Gates' minions instead. That's hard work, too, believe me!)

Supersymmetry all started, as far as we were concerned (and we were not aware of Golfand and Likhtman's work), with Julius Wess scurrying about secretively in his office during the autumn of 1973. He kept hiding away for hours on end with this good-looking Italian guy. Both of my subsequent wives have been great fans of Bruno's over the years, so I know what I am talking about! I was busy with other things at the time: studying for my *Diplom* exam was one, and getting married for the first time was another. Once both of these were duly out of the way, Julius (a) kindly agreed to keep me on as a Ph.D. student, and (b) announced an institute seminar talk by himself with the title "Super-gauge Symmetry in Four Dimensions (in English)," or some such. Little did I know that 27 years later the paper he was going to talk about would be one of the most widely quoted in the literature.

The next thing that happened was the arrival of a couple of preprints from Imperial College. Abdus Salam and John Strathdee had taken up the ball, started running with it, and had shown that Wess and Zumino's algebra was $SU(2,2|1)$, as well as invented superspace while they were at it. Personally, I was most impressed by their proposal to drop the "-gauge" from Wess and Zumino's baby and call it plain "super-symmetry" – with the hyphen. The "-gauge" bit had been a left-over from the two-dimensional conformal super-gauge symmetry (which has infinitely many generators) of old fermionic string theory. Thus, a new name was born! I don't know the current whereabouts of John, but I sure hope to see him and renew our acquaintance at the symposium. The last I heard, in the '80s, he was off to New Zealand to raise sheep, only coming back once in a while to write a paper with Salam. (Or was the sheep thing a joke?)

Abdus Salam, as we all know, is sadly not with us anymore. A lot has been written and said about him since his illness and much too early death in 1996, but that death is now long enough ago that I may be permitted to resurrect the image of the living man as well as gossip a little. Talk about two wives! In 1982, my first wife came to England to visit Judith and me. As it happened, Abdus was in town, and I had a short conversation alone with him in his office at Imperial. I mentioned with the self-confidence of relative youth that Eva was visiting us and cheerily – I thought – added: "you see, I too can get along with two wives." Whereupon Abdus's face broadened into the kind of all-over wrinkly smile only seen on the faces of bearded Punjabis of a certain age, and with a sparkle in his eyes he leaned forward, touched my arm, and said: "But, my friend, you don't sleep with both of them." Advantage Professor Salam!

Back to serious things. By the beginning of 1974, I don't think Professor Wess was quite sure what to do with me. As I had a fairly prestigious doctoral scholarship, and was married to a wife who earned a good salary, it was probably the right thing for him to give it a long shot, not actually bother about a thesis problem at all, but rather let me try my hand at primary research right away. Salam and Strathdee had hinted that it should be possible to extend $SU(2, 2|1)$ into $SU(2, 2|2)$. Julius told Peter Dondi, a postdoc recently arrived from DAMTP in Cambridge, to have a go at that and to let me in on the fun. Which he did. In consequence, I ended up with a case of Repetitive Strain Injury to my wrists from multiplying 6×6 matrices in all possible ways by means of pencil and paper, and with my first publication. It was the sixth in the new field of Supersymmetry (not counting Golfand and Likhtman, of whom we were blissfully ignorant, or Volkov and Akulov whose stuff we did not understand).

In early summer, Jan Lopuszański came to visit from Wroclaw for a few weeks before he was due to go on to visit CERN for the rest of the summer. Wess told him what he had been working on, only to draw the disbelieving comment from Jan that O'Raifeartaigh and others had shown that such a thing could not be done. Jan was a stubborn man, and I dont think that in those initial conversations he quite grasped at first the major "new thing," the anti-commutators. In the end, Julius became impatient, and – quite politely, of course, as behooves an Austrian of the old school when talking to a distinguished member of the Polish gentry – he told Jan to go away and "talk to Martin who has done a lot of this algebra stuff." My good luck!

Jan quickly changed tack. Obviously, the S-matrix guys (Sidney Coleman, Jeff Mandula, Lochlainn O'Raifeartaigh, and possibly others) hadn't been wrong; it was just that they had gone past the particular cross-road leading to fermions without paying much attention. They had squarely aimed at Pati, Salam, and Strathdee's earlier attempt to grandly unify particle symmetries with space-time, and had blown it out of the sky quite spectacularly – hence, of course, that immediate interest of the Imperial College group in "super-symmetry"! Jan set to work, with a bit of help from me, to examine what was left over of the older no-go theorems when taking the possibility of supersymmetry into account. Together, we produced a rather confused though almost certainly correct paper which, I believe, neither of us fully understood. (Well, let me be fair: all I can definitely say is that I, for one, did not understand the axiomatic field-theory bits of it.) The paper was submitted, and Jan left for CERN. Within a couple of weeks two surprising things happened: the paper was accepted unchanged by Nuclear Physics B almost by return of post, and Jan withdrew it. I was flabbergasted. What had happened?

Well, Jan had run into Rudolf Haag, on a sabbatical from Hamburg to CERN that year. Together, they had taken the thing, torn it into its constituent parts, thrown those up into the air, and had them land on the ground with a solid, satisfying thud: a self-consistent, beautifully reasoned, clear piece of work, hewn into granite and set in concrete. As Haag later once said to me when I was trying to slip an iffy calculation into the paper: "I have a reputation to lose." Bless him! Of course, the original Lopuszański-Sohnius paper had been shrunk to about half its size.

Nowadays, of course, one would make sure to add two or even three papers to one's list of publications in a situation like this. Not so in 1974. I was working with a pair of senior full professors, secure in their positions, to whom their professional pride counted far more than an extra notch on their already well-worn belts. So the original paper was withdrawn. Moreover, Haag was very interested at the time in the conformal-invariance aspects of Wess & Zumino. He wanted to extend the paper to include that, too. Being a master of analytic S-matrix theory, he also knew exactly what could and what could not survive of the no-go theorems in the zero-mass limit.

Let me slip in a vignette about Haag, the physicist. Ask anyone, and they'll probably say that physicists don't come much more mathematical than him. And yet: for a couple of years after 1974, he and I worked on and off on a paper to explore the consequences of conformal supersymmetry on the S-matrix. The idea was that there should be a lot of form-factors that result from the underlying symmetry – Clebsch-Gordan coefficients for $SU(2,2|n)$, so to speak. A lovely mathematical problem, and since this was a space-time symmetry, those "Clebsches" would make predictions about scattering angles and all sorts of wonderful things. Then, quite suddenly, Rudolf lost interest, and nothing ever happened to that work. (I published a tiny bit of it, with his permission, much later in Jan's 60th-birthday volume.) In the '80s, I asked Rudolf why he had so suddenly cooled off. He answered "because you told me that we had been wrong." Huh? I had told him we were wrong? When? How? "Well, it was you who pointed out to me that the zero-mass approximation meant that not only the electron mass had to be small against the scattering energy, but also the W-boson mass. And that feat even LEP will not be able to achieve, so why bother calculating scattering dynamics for a non-existing physical situation?" So, the great man had made a silly mistake, the Ph.D. student had always been aware of the mistake, but discarded it because he thought that life was long and that the maths was fun independent of the physics. But when he told the great man, the physicist in Haag won out over the mathematician and he lost interest totally and completely.

By the end of the summer 1974, Jan's time had run out – back home to the Workers' Paradise, where the first stirrings were faintly heard from the shipyards in Gdansk around that time! So, next thing, recent graduate Martin S. was put on a train to Geneva (my first business trip ever), picked up at the station by an elderly (about my age now) and never-before seen gentleman who chain-smoked, chewed on a blade of grass and rather slowly drove a completely clapped out estate-car ("station wagon" to some of you). He took me to his flat, and we spent the next week or so calculating and writing at the kitchen table for about 17 hours a day and sharing the bedroom in exhausted sleep for the remaining seven. (I didn't snore then and luckily neither did he – I wish he could let me into the secret of how to manage that as a man of over 50! Should I start chewing on blades of grass?)

Thus, in three easy stages, was conceived *Nucl. Phys.* **B88**, 275 (1975).

The summer was over, a new academic year about to begin, and there was a new kid on the block. In fact, the new kid had been on my block for some time – he was the kid from next door, literally, from number 23 to the Sohnius' number 21. For many years we had seen him walk past the house on his way from school, a gangly, somewhat shy but very obviously bright teenager. Our parents were colleagues and friends, occasional dinners, neighborly barbecues and glasses of wine – that sort of thing. He started reading physics at the local uni, just as I had done a couple of years earlier, and in due course had shown up in the exercise classes that went with Julius' lecture course on Theoretical Physics. Talk about a bright kid! Most of my readers will have taught undergraduates and be familiar with the kind of messy pieces of paper you often get back from them: sloppy argumentation, half-guesses, wild detours leading to the correct result, or not. Not so this guy. Straight to the bone, three tight lines of maths zeroing in on the correct result, each time. I could hardly believe it! Then he asked Julius to be his thesis adviser at just about the same time as our parents fell out over some domestic nonsense. Let's just say it was neither his nor his parents' fault. Consequently, stupidly, and utterly unprofessionally, I asked Wess to be allowed to stay at arm's length from the new kid. My bad luck! His name was, and is, Hermann Nicolai. Is that why that second volume of my Physics Report, the one on Supergravity, never got written? I don't know.

The rest, they say, is history ...

ON SUPERPARTNERS AND THE ORIGINS OF
SUPERSYMMETRIC STANDARD MODEL

P. FAYET

Laboratoire de Physique Théorique de l'Ecole Normale Supérieure,[b]
24 rue Lhomond, 75231 Paris Cedex 05, France.

We recall the obstacles which seemed, long ago, to prevent supersymmetry from possibly being a fundamental symmetry of Nature. Which bosons and fermions could be related? Is spontaneous supersymmetry breaking possible? Where is the spin-$\frac{1}{2}$ Goldstone fermion of supersymmetry? Can one define conserved baryon and lepton numbers in such theories, although they systematically involve self-conjugate Majorana fermions? etc.. We then recall how an early attempt to relate the photon with a "neutrino" led to the introduction of R-invariance, but that this "neutrino" had to be reinterpreted as a new particle, the *photino*. This led us to the Supersymmetric Standard Model, involving the SU(3)×SU(2)×U(1) gauge interactions of chiral quark and lepton superfields, and of two doublet Higgs superfields responsible for the electroweak breaking and the generation of quark and lepton masses. The original continuous R-invariance was then abandoned in favor of its discrete version, R-parity – reexpressed as $(-1)^{2S}\,(-1)^{(3B+L)}$ – so that the gravitino and gluinos can acquire masses. We also comment about supersymmetry breaking.

Introduction

The algebraic structure of supersymmetry in four dimensions was introduced in the beginning of the seventies by Gol'fand and Likhtman,[1] Volkov and Akulov,[2] and Wess and Zumino,[3] as recalled in various contributions to this book. It involves a spin-$\frac{1}{2}$ fermionic symmetry generator, called the supersymmetry generator, satisfying anticommutation relations. This supersymmetry generator Q is defined so as to relate fermionic with bosonic fields, in supersymmetric relativistic quantum field theories.

At that time it was not at all clear if – and even less how – supersymmetry could actually be used to relate fermions and bosons, in a physical theory of particles. While very interesting from the point of view of relativistic field theory, supersymmetry seemed clearly inap-

[b]UMR 8549, Unité Mixte du CNRS et de l'Ecole Normale Supérieure.

propriate for a description of our physical world. In particular one could not identify physical bosons and fermions that might be related under such a symmetry. It even seemed initially that supersymmetry could not be spontaneously broken at all – in contrast with ordinary symmetries – which would imply that bosons and fermions be systematically degenerated in mass! Supersymmetric theories also involve, systematically, self-conjugate Majorana spinors – unobserved in Nature – while the fermions that we know all appear as Dirac fermions carrying conserved (B and L) quantum numbers. In addition, how could we account for the conservation of the fermionic numbers B and L (only carried by fermions) in a supersymmetric theory in which fermions are related to bosons? Most physicists were then considering supersymmetry as irrelevant for "real physics."

Still, this algebraic structure could actually be taken seriously as a possible symmetry of physics of fundamental particles and interactions once we understood that the above obstacles preventing the application of supersymmetry to the real world could be overcome. After an initial attempt illustrating how far one could go in trying to relate known particles together (in particular, the photon with a "neutrino," and the $W^{\pm}$ bosons with charged "leptons" – and the limitations of this approach – in a spontaneously broken $SU(2) \times U(1)$ electroweak theory involving two chiral doublet Higgs superfields[4]) we were quickly led to reinterpret the fermions of this model, which all possess a conserved R quantum number carried by the supersymmetry generator, as belonging to a new class of particles. The "neutrino" ought to really be considered as a new particle, a "photonic neutrino," a name which I transformed in 1977 into *photino*, also calling at the same time *gluinos* the fermionic partners of the colored gluons (quite distinct from the quarks!), and so on. More generally, this led us to postulate the existence of new R-odd "superpartners" for all ordinary particles and consider them seriously, despite their rather non-conventional properties: e.g. new bosons carrying "fermion" number, now known as *sleptons* and *squarks*, or Majorana fermions transforming as an $SU(3)$ color octet, which are precisely the *gluinos*, etc.[5,6] In addition the electroweak breaking must be induced by *a pair* of electroweak Higgs doublets, not just a single one as in the Standard Model, which requires the existence of *charged*

Higgs bosons, and of several neutral ones.

The still-hypothetical superpartners may be distinguished by a new quantum number called R-parity,[7] associated with a Z_2 remnant of the continuous R-symmetry, which may be multiplicatively conserved in a natural way, and is especially useful to guarantee the absence of unwanted interactions mediated by squark or slepton exchanges. The conservation (or non-conservation) of R-parity is closely related with the conservation (or nonconservation) of baryon and lepton numbers, B and L, as illustrated by the well-known formula[8] reexpressing R-parity as $(-1)^{2S}(-1)^{3B+L}$. The finding of the basic building blocks of what we now call the Supersymmetric Standard Model (whether "minimal" or "nonminimal") allowed for the experimental searches for "supersymmetric particles," which started with the first searches for gluinos and photinos, selectrons and smuons, in the years 1978-1980, and have been going on continuously since. These searches often rely on the "missing energy" signature, corresponding to energy-momentum carried away by unobserved neutralinos.[5,8,9,10] A conserved R-parity also ensures the stability of the "lightest supersymmetric particle," a good candidate to constitute the nonbaryonic Dark Matter that seems to be present in our Universe. The general opinion of the scientific community towards supersymmetry and supersymmetric extensions of the Standard Model has considerably changed since the early days, and it is now widely admitted that supersymmetry may well be the next fundamental symmetry to be discovered in the physics of fundamental particles and interactions, although this remains to be experimentally proven.

Nature does not seem to be supersymmetric!

Let us now travel back in time, and think about the supersymmetry algebra, and the way it might be realized in Nature. This supersymmetry algebra

$$\left\{ \begin{array}{rcl} \{\, Q,\, \bar{Q}\,\} & = & -2\gamma_\mu P^\mu \quad, \\ [\, Q,\, P^\mu\,] & = & 0 \quad. \end{array} \right. \tag{1}$$

was introduced, in the years 1971-1973, by three different groups, with quite different motivations. Gol'fand and Likhtman,[1] in their remarkable work published in 1971, first introduced it with the apparent hope of understanding parity violation: when the Majorana supersymmetry generator Q_α is written as a two-component chiral Dirac spinor (say Q_L), one may have the impression that the supersymmetry algebra, which then involves a chiral projector in the right-hand side of the anticommutation relation (1), is intrinsically parity violating (which, however, is not the case); they suggested that such (supersymmetric) models must therefore necessarily violate parity, probably thinking that this could lead to an explanation for parity violation in weak interactions. Volkov and Akulov[2] hoped to explain the masslessness of the neutrino from a possible interpretation as a spin-$\frac{1}{2}$ Goldstone particle, while Wess and Zumino[3] wrote the algebra by extending to four dimensions the "supergauge" (i.e. supersymmetry) transformations,[11] and the algebra,[12] acting on the two-dimensional string world sheet. However, the mathematical existence of an algebraic structure does not imply that it has to play a rôle as an invariance of the fundamental laws of Nature.[c]

Indeed many obstacles seemed, long ago, to prevent supersymmetry from possibly being a fundamental symmetry of Nature. Which bosons and fermions could be related by supersymmetry? May be supersymmetry could act at the level of composite objects, e.g. as relating baryons with mesons? Or should it act at a fundamental level, i.e. at the level of quarks and gluons? (But quarks are color triplets, and electrically charged, while gluons transform as an SU(3) color

---

[c]Incidentally while supersymmetry is commonly referred to as "relating fermions with bosons," its algebra (1) does not even require the existence of fundamental bosons! (With nonlinear realizations of supersymmetry a fermionic field can be transformed into a *composite* bosonic field made of fermionic ones[2]; but we shall work within the framework of the linear realizations of the supersymmetry algebra, which allows for renormalizable supersymmetric field theories.) The supersymmetry algebra (1) certainly does not imply by itself the existence of the superpartners! (Just as the mathematical existence of the SU(2) group does not imply the physical existence of the isospin or electroweak symmetries, the existence of SU(3) does not imply that of the strange quark, and the flavor or color symmetries; the existence of SU(4) does not require technicolor, nor that of SU(5), ,grand unification!)

octet, and are electrically neutral!) Is spontaneous supersymmetry breaking possible at all? If yes, where is the spin-$\frac{1}{2}$ Goldstone fermion of supersymmetry if it cannot be identified as one of the known neutrinos? Can we use supersymmetry to relate directly known bosons and fermions? And, if not, why? If known bosons and fermions cannot be directly related by supersymmetry, do we have to accept this as the sign that supersymmetry is *not* a symmetry of the fundamental laws of Nature? If we still insist to work within the framework of supersymmetry, how could it be possible to define conserved baryon and lepton numbers in such theories, which systematically involve *self-conjugate* Majorana fermions, unknown in Nature, while B and L are carried only by fundamental (Dirac) fermions – not by bosons? And, once we are finally led to postulate the existence of new bosons carrying B and L – the new spin-0 squarks and sleptons – can we prevent them from mediating new unwanted interactions?

While bosons and fermions should have equal masses in a supersymmetric theory, this is certainly not the case in Nature. Supersymmetry should then clearly be broken. But spontaneous supersymmetry breaking is notoriously difficult to achieve, to the point that it was even initially thought to be impossible! Why is it so? Supersymmetry is a special symmetry, since the Hamiltonian, which appears in the right-hand side of the anticommutation relations (1), can be expressed proportionally to the sum of the squares of the components of the supersymmetry generator, as $H = \frac{1}{4} \sum_\alpha Q_\alpha^2$. This implies that a supersymmetry preserving vacuum state must have vanishing energy,[13] while any candidate for a "vacuum state" which would not be invariant under supersymmetry may naïvely be expected to have a larger, positive, energy.[d] As a result, potential candidates for supersymmetry breaking vacuum states seemed to be necessarily unstable, leading to the question:

Q1 :   *Is spontaneous supersymmetry breaking possible at all?* (2)

As it turned out, and despite the above argument, several ways of

---

[d]Such a would-be supersymmetry breaking state corresponds, in global supersymmetry, to a *strictly positive* energy density – the scalar potential being expressed proportionally to the sum of the squares of the auxiliary D, F and G components, as $V = \frac{1}{2} \sum (D^2 + F^2 + G^2)$.

breaking spontaneously global or local supersymmetry have been found.[14,1,16] But spontaneous supersymmetry breaking remains, in general, rather difficult to obtain, since theories tend to prefer, for energy reasons, supersymmetric vacuum states. Only in very exceptional situations can the existence of such states be completely avoided!

As explained above in global supersymmetry a non-supersymmetric state has, in principle, always more energy than a supersymmetric one; it then seems that it should always be unstable, the stable vacuum state being, necessarily, a supersymmetric one! Still it is possible to escape this general result – and this is the key to spontaneous supersymmetry breaking – if one can arrange to be in one of those rare situations for which *no supersymmetric state exists at all* – the set of equations for the auxiliary field VEV's $\langle D \rangle$'s $= \langle F \rangle$'s $= \langle G \rangle$'s $= 0$ having *no solution at all*. But these situations are in general quite exceptional. (This is in sharp contrast with ordinary symmetries, in particular gauge symmetries, for which one only has to arrange for nonsymmetric states to have less energy than symmetric ones in order to get spontaneous symmetry breaking.) These rare situations usually involve an Abelian U(1) gauge group,[14] allowing for a gauge-invariant linear "ξD" term to be included in the Lagrangian density[e]; and/or an appropriate set of chiral superfields with special superpotential interactions which must be very carefully chosen (so as to get "F-breaking"),[15] preferentially with the help of additional symmetries such as R-symmetries. In local supersymmetry,[17] which includes gravity, one also has to arrange, at the price of a very severe fine-tuning, for the energy density of the vacuum to vanish exactly,[16] or almost exactly, to an extremely good accuracy, so as not to generate an unacceptably large value of the cosmological constant Λ.

Whatever the mechanism of supersymmetry breaking, we have to get – if this is indeed possible – a physical world which looks like ours (which will precisely lead to postulate the existence of superpartners for all ordinary particles). Of course just accepting the possibility of

---

[e]Even in the presence of such a term, one frequently does not get a spontaneous breaking of the supersymmetry: one has to be very careful so as to avoid the presence of supersymmetry restoring vacuum states, which generally tend to exist.

explicit supersymmetry breaking without worrying too much about the origin of supersymmetry breaking terms, as is frequently done now, makes things much easier – but also at the price of introducing a large number of arbitrary parameters, coefficients of these supersymmetry breaking terms. In any case such terms must have their origin in a spontaneous supersymmetry breaking mechanism, if we want supersymmetry to play a fundamental role, especially if it is to be realized as a local fermionic gauge symmetry, as in the framework of supergravity theories. We shall come back to this question of supersymmetry breaking later. In between, we note that the spontaneous breaking of the global supersymmetry must in any case generate a massless spin-$\frac{1}{2}$ Goldstone particle, leading to the next question,

Q2 :   *Where is the spin-$\frac{1}{2}$ Goldstone fermion of supersymmetry ?*
$$(3)$$

Could it be[2] one of the known neutrinos? A first attempt at implementing this idea within a SU(2) × U(1) electroweak model of "leptons"[4] quickly illustrated that it could not be pursued very far. (Actually, the "leptons" of this first electroweak model were soon reinterpreted to become the "charginos" and "neutralinos" of the Supersymmetric Standard Model.)

If the Goldstone fermion of supersymmetry is not one of the known neutrinos, why hasn't it been observed? Today we tend not to think at all about the question, since: 1) the generalized use of soft terms breaking *explicitly* the supersymmetry seems to make this question irrelevant; 2) since supersymmetry has to be realized locally anyway, within the framework of supergravity,[17] the massless spin-$\frac{1}{2}$ Goldstone fermion ("goldstino") should in any case be eliminated in favor of extra degrees of freedom for a massive spin-$\frac{3}{2}$ gravitino.[6,16]

But where is the gravitino, and why has no one ever seen a fundamental spin-$\frac{3}{2}$ particle? Should this already be taken as an argument against supersymmetry and supergravity theories? Or should one consider that the crucial test of such theories should be the discovery of a spin-$\frac{3}{2}$ particle? In that case, how could it manifest its presence? In fact to discuss this question properly we need to know how this spin-$\frac{3}{2}$ particle should couple to the other particles, which requires us to know which bosons and fermions could be associated under supersymmetry.[5] In any case, even without knowing that, we might already anticipate that the interactions of the gravitino, being proportional to the square root of the Newton constant $\sqrt{G_N} \simeq 10^{-19}$ GeV^{-1}, should be absolutely negligible in particle physics experiments. Quite surprisingly, however, this may not necessarily be true! We might be in a situation for which the gravitino is light, maybe even extremely light, so that this spin-$\frac{3}{2}$ particle would still interact very much like the massless spin-$\frac{1}{2}$ Goldstone fermion of global supersymmetry, according to the "equivalence theorem" of supersymmetry.[6] In that case we are led back to our initial question, where is the spin-$\frac{1}{2}$ Goldstone fermion of supersymmetry? But at this point we are in a position to answer, the direct detectability of the gravitino depending crucially on the value of its mass $m_{3/2}$, which is fixed[6,18] by the supersymmetry breaking scale $\sqrt{d} = \Lambda_{ss}$.

In any case, much before getting to the Supersymmetric Standard Model, and irrespective of the question of supersymmetry breaking, the crucial question, if supersymmetry is to be relevant in particle physics, is:

$$\text{Q3}: \quad \textit{Which bosons and fermions could be related} \atop \textit{by supersymmetry?} \qquad (4)$$

But there seems to be no answer since known bosons and fermions do not appear to have much in common – excepted, maybe, for the photon and the neutrino. This track deserved to be explored,[4] but one cannot really go very far in this direction. In a more general way the number of (known) degrees of freedom is significantly larger for the fermions (now 90, for three families of quarks and leptons) than for the bosons (27 for the gluons, the photon and the $W^{\pm}$ and Z gauge bosons, ignoring for the moment the spin-2 graviton, and the still-undiscovered Higgs boson). And these fermions and bosons have very different gauge symmetry properties!

Furthermore supersymmetric theories also involve, systematically, self-conjugate Majorana spinors – unobserved in Nature – while the fermions that we know all appear as Dirac fermions carrying conserved B and L quantum numbers. This leads to the question

Q4 : *How could one define (conserved)*
baryon and lepton numbers, in a supersymmetric theory?
(5)

These quantum numbers, presently known to be carried by fundamental fermions only, not by bosons, seem to appear in Nature as *intrinsically-fermionic* numbers. Such a feature cannot be maintained in a supersymmetric theory, and one has to accept the (then rather heretic) idea of attributing baryon and lepton numbers to fundamental bosons, as well as to fermions. These new bosons carrying B or L are the superpartners of the spin-$\frac{1}{2}$ quarks and leptons, namely the now-familiar (although still unobserved) spin-0 *squarks* and *sleptons*. Altogether, all known particles should be associated with new *superpartners*.[5]

Of course nowadays we are so used to dealing with spin-0 squarks and sleptons, carrying baryon and lepton numbers almost by definition, that we can hardly imagine this could once have appeared as a problem. Its solution went through the acceptance of the idea of attributing baryon or lepton numbers to a large number of new fundamental bosons. But if such new spin-0 squarks and sleptons are introduced, their direct (Yukawa) exchanges between ordinary quarks and leptons, if allowed, could lead to an immediate disaster, preventing us from getting a theory of electroweak and strong interactions mediated by spin-1 gauge bosons, and not spin-0 particles, with conserved B and L quantum numbers! This may be expressed by the following question

Q5 : *How can we avoid unwanted interactions*
mediated by spin-0 squark and slepton exchanges?
(6)

Fortunately, we can naturally avoid such unwanted interactions, thanks to R-parity (a remnant of the continuous U(1) R-symmetry) which, if present, guarantees that squarks and sleptons can*not* be directly exchanged between ordinary quarks and leptons, allowing for conserved baryon and lepton numbers in supersymmetric theories.

Continuous R-invariance and electroweak symmetry breaking (from an early attempt to relate the photon and the neutrino)

Let us now return to an early attempt at relating *existing* bosons and fermions together,[4f] also at the origin of the definition of the continuous R-invariance (the discrete version of which leading to R-parity). It also showed how one can break spontaneously the $SU(2) \times U(1)$ electroweak gauge symmetry in a supersymmetric theory, using (in a modern language) a pair of chiral doublet Higgs superfields that would now be called H_1 and H_2. This involves a mixing angle initially called δ but now known as β, defined by

$$\tan \beta = \frac{v_2}{v_1}. \tag{7}$$

The fermions of this early supersymmetric model, which are in fact gaugino-higgsino mixtures, should no longer be considered as lepton candidates, but became essentially the "charginos" and "neutralinos" of the Supersymmetric Standard Model.[5,6]

Despite the general lack of similarities between known bosons and fermions, we tried as an exercise to see how far one could go in attempting to relate the spin-1 photon with a spin-$\frac{1}{2}$ neutrino. If we want to attempt to identify the companion of the photon as being a "neutrino," despite the fact that it initially appears as a self-conjugate Majorana fermion, we need to understand how this particle could carry a conserved quantum number that we might interpret as a "lepton" number. This was made possible by the introduction of *a continuous $U(1)$ R-invariance*,[4] which also guaranteed the mass-lessness of this "neutrino" ("ν_L" carrying $+1$ unit of R), by acting chirally on the Grassmann coordinate θ which appears in the expression of the various gauge and chiral superfields. The supersymmetry generator Q_α carries one unit of the corresponding additive conserved quantum number, called R (so that one has $\Delta R = \pm 1$ between a boson and a fermion related by supersymmetry).

---

[f] This model is reminiscent of a presupersymmetry two-Higgs-doublet model [19] which turned out to be very similar to supersymmetric gauge theories, with Yukawa and φ^4 interactions restricted by a continuous Q-invariance, ancestor of the continuous R-invariance of supersymmetric theories.

Attempting to relate the photon with one of the neutrinos could only be an exercise of limited validity. The would-be "neutrino," in particular, while having in this model a $V - A$ coupling to its associated "lepton" and the charged $W^{\pm}$ boson, was in fact what we would now call a "photino," not directly coupled to the Z boson! Still this first attempt, which essentially became a part of the Supersymmetric Standard Model, illustrated how one can break spontaneously a $SU(2) \times U(1)$ gauge symmetry in a supersymmetric theory through an electroweak breaking induced by *a pair of chiral doublet Higgs superfields*, now known as H_1 and H_2! (Using only a single doublet Higgs superfield would have left us with a massless charged chiral fermion, which is, evidently, unacceptable.) Our previous charged "leptons" were in fact what we now call two winos, or charginos, obtained from the mixing of charged gaugino and higgsino components, as given by the mass matrix

$$
\mathcal{M} \;=\; \begin{pmatrix} (\,m_2 = 0\,) & \dfrac{g\,v_2}{\sqrt{2}} = m_W \sqrt{2}\,\sin\beta \\[2ex] \dfrac{g\,v_1}{\sqrt{2}} = m_W \sqrt{2}\,\cos\beta & \mu = 0 \end{pmatrix}\;, \quad (8)
$$

in the absence of a direct higgsino mass that would have originated from a $\mu H_1 H_2$ mass term in the superpotential.[9] The whole construction showed that one could deal elegantly with elementary spin-0 Higgs fields (not a very popular ingredient at the time), in the framework of spontaneously-broken supersymmetric theories. Quartic Higgs couplings are no longer completely arbitrary, but fixed by the values of the gauge coupling constants – here the electroweak couplings g and g' – through the following "D-terms" (i.e. $\frac{\vec{D}^2}{2} + \frac{D'^2}{2}$)

---

[9]This $\mu H_1 H_2$ term, which would have broken explicitly the continuous $U(1)$ R-invariance then intended to be associated with the "lepton" number conservation law, was already replaced by a $\lambda H_1 H_2 N$ trilinear coupling involving an *extra neutral singlet chiral superfield* N.

in the scalar potential given in[4h]

$$V_{\text{Higgs}} = \frac{g^2}{8} \, (\, h_1^{\dagger} \, \vec{\tau} \, h_1 + h_2^{\dagger} \, \vec{\tau} \, h_2 \,)^2 + \frac{g'^2}{8} \, (\, h_1^{\dagger} \, h_1 - h_2^{\dagger} \, h_2 \,)^2 + \ldots$$

$$= \frac{g^2 + g'^2}{8} \, (\, h_1^{\dagger} \, h_1 - h_2^{\dagger} \, h_2 \,)^2 + \frac{g^2}{2} \, | \, h_1^{\dagger} \, h_2 \, |^2 + \ldots .$$

$$(9)$$

This is precisely the quartic Higgs potential of the "minimal" version of the Supersymmetric Standard Model, the so-called MSSM, with its quartic Higgs coupling constants equal to

$$\frac{g^2 + g'^2}{8} \quad \text{and} \quad \frac{g^2}{2} \, . \tag{10}$$

Further contributions to this quartic Higgs potential also appear in the presence of additional superfields, such as the neutral singlet chiral superfield N already introduced in this model, which will play an important rôle in the NMSSM, i.e. in "next-to-minimal" or "non-minimal" versions of the Supersymmetric Standard Model. Charged Higgs bosons (now called $H^{\pm}$) are present in this framework, as well as several neutral ones. Their mass spectrum depends on the details of the supersymmetry breaking mechanism considered: soft breaking terms, possibly "derived from supergravity," presence or absence of extra-U(1) gauge fields and/or additional chiral superfields, rôle of radiative corrections, etc.

The Supersymmetric Standard Model

These two Higgs doublets are precisely the two doublets used in 1977 to generate the masses of charged leptons and down quarks, and of up quarks, in supersymmetric extensions of the standard model.[5] Note that at the time having to introduce Higgs fields was generally considered as rather unpleasant. While one Higgs doublet was taken as probably unavoidable to get to the standard model or at least simulate the effects of the spontaneous breaking of the electroweak symmetry, having to consider two Higgs doublets, necessitating charged

---

[h]With a different denomination for the two Higgs doublets, such that $\varphi'' \mapsto h_1$, $(\varphi')^c \mapsto h_2$, $\tan \delta = v'/v'' \mapsto \tan \beta = v_2/v_1$.

Higgs bosons as well as several neutral ones, was usually considered as too heavy a price, in addition to the "doubling of the number of particles", once considered as an indication of the irrelevance of supersymmetry. As a matter of fact considerable work was devoted for a time on attempts to avoid fundamental spin-0 Higgs fields, before returning to fundamental Higgses, precisely in this framework of supersymmetry.

In the previous SU(2)×U(1) model,[4] it was impossible to view seriously for very long "gaugino" and "higgsino" fields as possible building blocks for our familiar lepton fields. This led us to consider that all quarks and leptons ought to be associated with new bosonic partners, the *spin-0 squarks and sleptons*. Gauginos and higgsinos, mixed together by the spontaneous breaking of the electroweak symmetry, correspond to a new class of fermions, now known as "charginos" and "neutralinos". In particular, the partner of the photon under supersymmetry, which cannot be identified with any of the known neutrinos, should be viewed as a new "photonic neutrino," the *photino*; the fermionic partner of the gluon octet is an octet of self-conjugate Majorana fermions called *gluinos*, etc. – although at the time *colored fermions* belonging to *octet* representations of the color SU(3) gauge group were generally believed not to exist (to the point that one could think of using the absence of such particles as a general constraint to select admissible grand-unified theories.[20])

The two doublet Higgs superfields H_1 and H_2 generate quark and lepton masses[5i] in the usual way, through the familiar trilinear superpotential

$$W = h_e\, H_1 . \bar{E}\, L + h_d\, H_1 . \bar{D}\, Q - h_u\, H_2 . \bar{U}\, Q \ . \qquad (11)$$

---

[i]The correspondence between earlier notations for doublet Higgs superfields, and modern ones, is as follows:

$$S = \begin{pmatrix} S^0 \\ S^- \end{pmatrix} \text{ and } T = \begin{pmatrix} T^0 \\ T^- \end{pmatrix} \longmapsto H_1 = \begin{pmatrix} H_1^0 \\ H_1^- \end{pmatrix} \text{ and } H_2 = \begin{pmatrix} H_2^+ \\ H_2^0 \end{pmatrix}.$$

(left-handed)   (right-handed)   (both left-handed)

Furthermore, we originally denoted, generically, by S_i, left-handed, and T_j, right-handed, the chiral superfields describing the left-handed and right-handed spin-$\frac{1}{2}$ quark and lepton fields, together with their spin-0 partners.

Table 2:   The basic ingredients of the Supersymmetric Standard Model.

---

1) the three $SU(3) \times SU(2) \times U(1)$ gauge superfield representations;

2) the chiral quark and lepton superfields corresponding to the three quark and lepton families;

3) the two doublet Higgs superfields H_1 and H_2 responsible for the spontaneous electroweak symmetry breaking, and the generation of quark and lepton masses through

4)   the trilinear superpotential (11) .

L and Q denote the left-handed doublet lepton and quark super-fields, and $\bar{E}$, $\bar{D}$ and $\bar{U}$ left-handed singlet antilepton and antiquark superfields. The vacuum expectation values of the two Higgs doublets described by H_1 and H_2 generate charged-lepton and down-quark masses, and up-quark masses, given by $m_e = h_e v_1/2$, $m_d = h_d v_1/2$ and $m_u = h_u v_2/2$, respectively. This constitutes the basic structure of the **Supersymmetric Standard Model**, which involves, at least, the ingredients shown in Table 2. Other ingredients, such as a direct $\mu\, H_1 H_2$ direct mass term in the superpotential, or an extra singlet chiral superfield N with a trilinear superpotential coupling $\lambda H_1 H_2 N + ...$ possibly acting as a replacement for a direct $\mu\, H_1 H_2$ mass term,[4] and/or extra U(1) factors in the gauge group (which could have been responsible for spontaneous supersymmetry breaking) may or may not be present, depending on the particular version of the Supersymmetric Standard Model considered.

In any case, independently of the details of the supersymmetry breaking mechanism ultimately considered, we obtain the following minimal particle content of the Supersymmetric Standard Model,

Table 3:   Minimal particle content of the Supersymmetric Standard Model.

Spin 1	Spin 1/2	Spin 0
gluons g photon γ	gluinos $\tilde{g}$ photino $\tilde{\gamma}$	
$W^{\pm}$ Z	winos $\widetilde{W}^{\pm}_{1,2}$ zinos $\widetilde{Z}_{1,2}$ higgsino $\tilde{h}^0$	$\left. \begin{array}{l} H^{\pm} \\[4pt] H \\[12pt] h,\ A \end{array} \right\}$ Higgs bosons
	leptons l quarks q	sleptons $\tilde{l}$ squarks $\tilde{q}$

given in Table 3. Each spin-$\frac{1}{2}$ quark q or charged lepton l^- is associated with *two* spin-0 partners collectively denoted by $\tilde{q}$ or $\tilde{l}^-$, while a left-handed neutrino ν_L is associated with a *single* spin-0 sneutrino $\tilde{\nu}$. We have ignored for simplicity further mixings between the various "neutralinos" described by neutral gaugino and higgsino fields, denoted in this table by $\tilde{\gamma}$, $\tilde{Z}_{1,2}$, and $\tilde{h}^0$. More precisely, all such models include four neutral Majorana fermions at least, corresponding to mixings of the fermionic partners of the two neutral $SU(2) \times U(1)$ gauge bosons (usually denoted by $\tilde{Z}$ and $\tilde{\gamma}$, or $\tilde{W}_3$ and $\tilde{B}$) and of the two neutral higgsino components ($\tilde{h}^0_1$ and $\tilde{h}^0_2$). Nonminimal models also involve additional gauginos and/or higgsinos.

On supersymmetry breaking, and the way from R-invariance to R-parity

Let us now return to the definition of the continuous R-symmetry and discrete R-parity transformations. R-parity is associated with a Z_2 subgroup of the group of continuous U(1) R-symmetry transformations, acting on the gauge superfields and the two doublet Higgs superfields H_1 and H_2 as in Ref. 4, with their definition extended to quark and lepton superfields so that quarks and leptons carry $R = 0$, and squarks and sleptons, $R = \pm 1$ (more precisely, $R = +1$ for $\tilde{q}_L$, $\tilde{l}_L$, and $R = -1$ for $\tilde{q}_R$, $\tilde{l}_R$).[5] As we shall see later, R-parity appears in fact as the remnant of this continuous R-invariance when gravitational interactions are introduced,[6] in the framework of local supersymmetry (supergravity). Either the continuous R-invariance, or simply its discrete version of R-parity, if imposed, naturally forbid the unwanted direct exchanges of the new squarks and sleptons between ordinary quarks and leptons.

These continuous U(1) R-symmetry transformations, which act chirally on the anticommuting Grassmann coordinate θ appearing in the definition of superspace and superfields, act on the gauge and chiral superfields of the Supersymmetric Standard Model as follows:

$$
\left\{
\begin{array}{l}
V(x, \theta, \bar{\theta}) \longrightarrow V(x, \theta\, e^{-i\alpha}, \bar{\theta}\, e^{i\alpha}) \\
\qquad \text{for the } SU(3) \times SU(2) \times U(1) \\
\qquad \text{gauge superfields} \\[2mm]
H_{1,2}(x, \theta) \longrightarrow H_{1,2}(x, \theta\, e^{-i\alpha}) \\
\qquad \text{for the left-handed doublet Higgs} \\
\qquad \text{superfields } H_1 \text{ and } H_2 \\[2mm]
S(x, \theta) \longrightarrow e^{i\alpha}\, S(x, \theta\, e^{-i\alpha}) \\
\qquad \text{for the left-handed (anti)quark} \\
\qquad \text{and lepton superfields} \\
\qquad Q, \bar{U}, \bar{D}, L, \bar{E} .
\end{array}
\right.
\tag{12}
$$

They are defined so as not to act on ordinary particles, which have $R = 0$, while their superpartners have, therefore, $R = \pm 1$. They allow us to distinguish between two separate sectors of R-even and R-odd particles. R-even particles include the gluons, photon, $W^{\pm}$ and Z gauge bosons, the various Higgs bosons, the quarks and leptons – and the graviton. R-odd particles include their superpartners, i.e. the gluinos and the various neutralinos and charginos, squarks and sleptons – and the gravitino (cf. Table 4). According to this first definition, R-parity simply appears as the parity of the additive quantum number R, as given[7] by the following expression:

$$R\text{-parity } R_p = (-1)^R = \begin{cases} +1 & \text{for ordinary particles,} \\ -1 & \text{for their superpartners.} \end{cases}$$
(13)

But why should we limit ourselves to the discrete R-parity symmetry, rather than considering its full continuous parent R-invariance? This *continuous* U(1) R-invariance, from which R-parity has emerged, is indeed a symmetry[5] of all four necessary basic building blocks of the Supersymmetric Standard Model:

1) the Lagrangian density for the SU(3)×SU(2)×U(1) gauge superfields;

2) the SU(3)×SU(2)×U(1) gauge interactions of the quark and lepton superfields;

3) the SU(2)×U(1) gauge interactions of the two chiral doublet Higgs superfields H_1 and H_2 responsible for the electroweak symmetry breaking;

4) and the trilinear "super-Yukawa" interactions (11) responsible for quark and lepton masses. Indeed this trilinear superpotential transforms under the continuous R-symmetry (12) with "R-weight" $n_{\mathcal{W}} = \sum_i n_i = 2$, i.e. according to

$$\mathcal{W}(x, \theta) \;\rightarrow\; e^{2i\alpha}\, \mathcal{W}(x, \theta e^{-i\alpha}) \;;$$
(14)

its auxiliary "F-component" (obtained from the coefficient of the bilinear $\theta\theta$ term in the expansion of $\mathcal{W}$), is therefore R-invariant, generating R-invariant interaction terms in the Lagrangian density.[j]

However, an unbroken continuous R-invariance, which acts chirally on the Majorana octet of gluinos,

$$\tilde{g} \;\rightarrow\; e^{\gamma_5 \alpha}\, \tilde{g}\,.$$
(15)

would constrain them to remain massless, even after a (spontaneous) breaking of the supersymmetry. We would then expect the existence

---

[j]Note, however, that a direct Higgs superfield mass term $\mu H_1 H_2$ in the superpotential, which has R-weight $n = 0$, does *not* lead to interactions which are invariant under the continuous R symmetry; but it gets in general reallowed, as for example in the MSSM, as soon as the continuous R symmetry gets reduced to its discrete version of R-parity.

Table 4: *R*-parities in the Supersymmetric Standard Model.

Bosons	Fermions
gauge and Higgs bosons ($R = 0$) graviton *R*-parity +	gauginos and higgsinos ($R = \pm 1$) gravitino *R*-parity −
sleptons ($R = \pm 1$) and squarks *R*-parity −	leptons ($R = 0$) and quarks *R*-parity +

of relatively light "*R*-hadrons"[8,9] made of quarks, antiquarks and gluinos, which have not been observed. In fact we know today that gluinos, if they do exist, should be rather heavy, requiring a significant breaking of the continuous *R*-invariance, in addition to the necessary breaking of the supersymmetry. Once the continuous *R*-invariance is abandoned, and supersymmetry is spontaneously broken, radiative corrections do indeed allow for the generation of gluino masses,[21] a point to which we shall return later.

Furthermore, the necessity of generating a mass for the Majorana spin-$\frac{3}{2}$ *gravitino*, once *local* supersymmetry is spontaneously broken, also forces us to abandon the continuous *R*-invariance, in favor of the discrete *R*-parity symmetry, thereby also allowing for gluino and other gaugino masses, at the same time as the gravitino mass $m_{3/2}$, as already noted in 1977[6]. (A third reason for abandoning the continuous *R*-symmetry could now be the non-observation at LEP of a charged *wino* – also called *chargino* – lighter than the $W^{\pm}$, that would exist in the case of a continuous U(1) *R*-invariance,[4,5] as shown by the mass matrix $\mathcal{M}$ of Eq. (8); the just-discovered τ^{-} particle could tentatively be considered, in 1976, as a possible light wino/chargino candidate, before it got clearly identified as a sequential heavy lepton.)

Once we drop the continuous R-invariance in favor of its discrete R-parity version, we may ask how general is this notion of R-parity, and, correlatively, are we *forced* to have this R-parity conserved? As a matter of fact, there is from the beginning a close connection between R-parity and baryon and lepton number conservation laws, which has its origin in our desire to get supersymmetric theories in which B and L could be conserved, and, at the same time, to avoid unwanted exchanges of spin-0 squarks and sleptons. Actually the superpotential of the theories discussed in Ref. 5 was constrained from the beginning, for that purpose, to be an *even* function of the quark and lepton superfields. *Odd* superpotential terms, which would have violated the "matter-parity" symmetry $(-1)^{(3B+L)}$, were excluded, to be able to recover B and L conservation laws, and avoid direct Yukawa exchanges of spin-0 squarks and sleptons between ordinary quarks and leptons. Tolerating unnecessary superpotential terms which are *odd* functions of the quark and lepton superfields (i.e. R_p-violating terms), does create, in general, immediate problems with baryon and lepton number conservation laws (most notably, a much too fast proton instability, if both B and L violations are simultaneously allowed).

This intimate connection between R-parity and B and L conservation laws can be made quite obvious by noting that for usual particles $(-1)^{2S}$ coincides with $(-1)^{3B+L}$, so that the R-parity (13) may be re-expressed[8] in terms of the spin S and the "matter-parity" $(-1)^{3B+L}$, as follows:

$$R\text{-parity} \; = \; (-1)^{\,2S}\,(-1)^{\,3B+L} \; . \qquad (16)$$

This may also be written as $(-1)^{2S}(-1)^{3(B-L)}$, showing that this discrete symmetry may still be conserved even if baryon and lepton numbers are separately violated, as long as their difference $(B-L)$ remains conserved, at least modulo 2.

The R-parity symmetry operator may also be viewed as a nontrivial geometrical discrete symmetry associated with a reflection of the anticommuting fermionic Grassmann coordinate, $\theta \; \rightarrow \; -\theta$, in superspace.[22] This R-parity operator plays an essential rôle in the discussion of the experimental signatures of the new particles. A conserved R-parity guarantees that *the new spin-0 squarks and sleptons cannot be directly exchanged* between ordinary quarks and leptons, as well as the absolute stability of the "lightest supersymmetric particle" (or LSP), a good candidate for non-baryonic Dark Matter in the Universe.

Let us come back to the question of supersymmetry breaking. which still has not received a definitive answer yet. The inclusion, in the Lagrangian density, of universal soft supersymmetry breaking terms for all squarks and sleptons,

$$- \sum_{\tilde{q}, \tilde{l}} m_0^2 \ (\tilde{q}^\dagger \tilde{q} + \tilde{l}^\dagger \tilde{l}), \qquad (17)$$

was already considered in 1976. But it was also understood that such terms should in fact be generated by a spontaneous supersymmetry breaking mechanism, especially if supersymmetry is to be realized locally. As a matter of fact they were first spontaneously generated with the help of the "D-term" associated with an *extra* U(1) gauge symmetry, acting *axially* on leptons and quarks fields, thereby allowing to lift, by the same positive amount, the mass^2 of *both* "left-handed" and "right-handed" slepton and squark fields. When the gauge coupling constant g" of this (still unbroken) extra U(1) was taken to be very small, the supersymmetry was spontaneously broken "at a very high scale" $\sqrt{d} = \Lambda_{ss} \gg m_W$. In the limit g" $\to 0$, the corresponding Goldstone fermion, the gaugino of the extra U(1), became completely decoupled, but supersymmetry was still broken with heavy slepton and squark masses; the breaking was then explicit instead of spontaneous, although only softly through the dimension 2 mass terms (17).

To get a true spontaneous breaking of the supersymmetry with a physically coupled goldstino (of course to be subsequently "eaten" by the spin-$\frac{3}{2}$ gravitino) rather than an explicit (although soft) one, as well as a spontaneous breaking of the extra $U(1)$ symmetry, and also to render, at the same time, the superpotential (11) invariant under this extra $U(1)$ symmetry so that it can actually be responsible for the generation of lepton and quark masses, we modified the definition of this extra $U(1)$ so that it also acts on the Higgs superfields H_1 an H_2 as well as on lepton and quark superfields, as follows:

$$
\left\{
\begin{array}{l}
V(x,\,\theta,\,\bar{\theta}\,) \quad \longrightarrow \quad V(x,\,\theta,\,\bar{\theta}\,) \\
\qquad \text{for the } SU(3) \times SU(2) \times U(1) \\
\qquad \text{gauge superfields;} \\[2mm]
H_{1,2}(x,\,\theta) \quad \longrightarrow \quad e^{-i\alpha}\, H_{1,2}(x,\,\theta) \\
\qquad \text{for the left-handed doublet Higgs} \\
\qquad \text{superfields} \quad H_1 \text{ and } H_2 \\[2mm]
S(x,\,\theta) \quad \longrightarrow \quad e^{i\frac{\alpha}{2}}\, S(x,\,\theta) \\
\qquad \text{for the left-handed (anti)quark} \\
\qquad \text{and (anti)lepton superfields} \\
\qquad\qquad Q,\,\bar{U},\,\bar{D},\,L,\,\bar{E}\ .
\end{array}
\right.
\tag{18}
$$

This newly-defined extra U(1) (acting on the two Higgs doublets so that it gets spontaneously broken together with the electroweak symmetry), is a new symmetry of the trilinear superpotential interactions (11), so that lepton and quark can now acquire masses in a way compatible with the spontaneous supersymmetry breaking mechanism used. This extra U(1) is associated, in the simplest case of Eq. (18), with a purely axial extra U(1) current for all quarks and charged leptons. Gauging such an extra U(1), which must in any case be different from the weak-hypercharge U(1), is in fact necessary,[23] if one intends to generate large *positive* mass^2 for *all* squarks ($\tilde{u}_L$, $\tilde{u}_R$, $\tilde{d}_L$, $\tilde{d}_R$) and sleptons, at the classical level, in a spontaneously-broken globally supersymmetric theory (otherwise we could not avoid squarks having negative or at best very small mass^2). But this method of spontaneous supersymmetry breaking also led to several difficulties. In addition to the question of anomalies, it required new neutral current interactions beyond those of the Standard Model. This was fine at the time, in 1977, but such interactions did not show up while the SU(2)×U(1) neutral current structure of the Standard Model got experimentally confirmed. This mechanism also left us with the question of generating large gluino masses. The gauging of an extra U(1) no longer appears as an appropriate way to generate large superpartner masses. One now uses again, in general, soft supersymmetry-breaking terms[24] generalizing those of Eq. (17)

– possibly "induced by supergravity" – which essentially serve as a parametrization of our ignorance about the true mechanism of supersymmetry breaking chosen by Nature to make superpartners heavy.

Let us return to gluino masses. As we said before continuous R-symmetry transformations act *chirally* on gluinos, so that an unbroken R-invariance would require them to remain massless, even after a spontaneous breaking of the supersymmetry! Thus the need, once it became experimentally clear that massless or even light gluinos could not be tolerated, to generate a gluino mass either from radiative corrections,[21] or from supergravity (see already in ([6]), with, in both cases, the continuous R-invariance reduced to its discrete R-parity subgroup.

In the framework of global supersymmetry it is not so easy to generate large gluino masses. Even if global supersymmetry is spontaneously broken, and if the continuous R-symmetry is not present, it is still in general rather difficult to obtain large masses for gluinos, since: **i)** no direct gluino mass term is present in the Lagrangian density; and **ii)** no such term may be generated spontaneously, at the tree approximation, gluino couplings involving *colored* spin-0 fields. A gluino mass may then be generated by radiative corrections involving a new sector of quarks sensitive to the source of supersymmetry breaking,[21] that would now be called "messenger quarks," but **iii)** this can only be through diagrams which "know" both about: **a)** the spontaneous breaking of the global supersymmetry, through some appropriately-generated VEV's for auxiliary components, $\langle D \rangle$, $\langle F \rangle$ or $\langle G \rangle$'s; **b)** the existence of superpotential interactions which do not preserve the continuous U(1) R-symmetry. Such radiatively-generated gluino masses, however, generally tend to be rather small, unless one introduces, in some often rather complicated "hidden sector," very large mass scales $\gg m_W$.

Fortunately gluino masses may also result directly from supergravity, as already observed in 1977.[6] Gravitational interactions require, within local supersymmetry, that the spin-2 graviton be associated with a spin-3/2 partner,[17] the gravitino. Since the gravitino is the fermionic gauge particle of supersymmetry it must acquire a mass, $m_{3/2}(= \kappa d/\sqrt{6} \approx d/m_{\text{Planck}})$, as soon as the local supersymmetry gets spontaneously broken. Since the gravitino is a self-conjugate Majorana fermion its mass breaks the continuous R-invariance which acts chirally on it, just as for the gluinos, forcing us to abandon the continuous U(1) R-invariance, in favor of its discrete R-parity subgroup. In particular, in the presence of a spin-$\frac{3}{2}$ gravitino mass term $m_{3/2}$, which corresponds to $\Delta R = \pm 2$, the "left-handed sfermions" $\tilde{f}_L$, which carry $R = +1$, can mix with the right-handed" ones $\tilde{f}_R$, carrying $R = -1$, through mixing terms having $\Delta R = \pm 2$, which may naturally (but not necessarily) be of order $m_{3/2} m_f$. Supergravity theories offer, in addition, a natural framework in which to include direct gaugino Majorana mass terms

$$-\frac{i}{2} m_3 \, \bar{\tilde{G}}_a \tilde{G}_a \; - \frac{i}{2} \, m_2 \, \bar{\tilde{W}}_a \tilde{W}_a \; - \frac{i}{2} \, m_1 \, \bar{\tilde{B}} \, \tilde{B} \, , \qquad (19)$$

which also correspond to $\Delta R = \pm 2$. The mass parameters m_3, m_2 and m_1, for the SU(3×SU(2)×U(1) gauginos, could naturally (but not necessarily) be of the same order as the gravitino mass $m_{3/2}$. Incidentally, once the continuous R-invariance is reduced to its discrete R-parity subgroup, a direct Higgs superfield mass term $\mu H_1 H_2$, which was not allowed by the continuous U(1) R-symmetry, gets re-allowed in the superpotential, as for example in the MSSM. The size of this supersymmetric μ parameter (which breaks explicitly both the continuous R-invariance (12) and the (global) extra U(1) symmetry (18) may then be controlled by considering one or the other of these two symmetries. In general, irrespective of the supersymmetry breaking mechanism considered, one normally expects the various superpartners not to be too heavy, otherwise the corresponding new mass scale would tend to contaminate the electroweak scale, thereby *creating* a hierarchy problem in the Supersymmetric Standard Model. Superpartner masses are then normally expected to be naturally of the order of m_W, or at most in the $\sim$ TeV/c^2 range.

The Supersymmetric Standard Model ("minimal" or not), with its R-parity symmetry (absolutely conserved, or not), provided the basis for the experimental searches for the new superpartners and Higgs bosons, starting with the first searches for gluinos and photinos, selectrons and smuons, at the end of the seventies. How the supersymmetry should actually be broken, if indeed it is a symmetry of Nature, is not known yet. Many good reasons to work on the Supersymmetric Standard Model and its various extensions have been discussed, dealing with supergravity, the high-energy unification of the gauge couplings, extended supersymmetry, new spacetime dimensions, superstrings, "M-theory", ... However, despite all the efforts made for more than twenty years to discover the new inos and sparticles, we are still waiting for experiments to disclose the missing half of the SuperWorld!

References

1. Yu. A. Gol'fand and E.P. Likhtman, *ZhETF Pis. Red.* **13**, 452 (1971) [*JETP Lett.* **13**, 323 (1971)].
2. D.V. Volkov and V.P. Akulov, *Phys. Lett.* B **46**, 109 (1973).
3. J. Wess and B. Zumino, *Nucl. Phys.* B **70**, 39 (1974); *Phys. Lett.* B **49**, 52 (1974); *Nucl. Phys.* B **78**, 1 (1974).
4. P. Fayet, *Nucl. Phys.* B **90**, 104 (1975).
5. P. Fayet, *Phys. Lett.* B **64**, 159 (1976); B **69**, 489 (1977).
6. P. Fayet, *Phys. Lett.* B **70**, 461 (1977).
7. P. Fayet, in *New Frontiers in High-Energy Physics*, Proc. Orbis Scientiae, Coral Gables (Florida, USA), 1978, eds. A. Perlmutter and L.F. Scott (Plenum, N.Y., 1978) p. 413.
8. G.R. Farrar and P. Fayet, *Phys. Lett.* B **76**, 575 (1978).
9. G.R. Farrar and P. Fayet, *Phys. Lett.* B **79**, 442 (1978).
10. G.R. Farrar and P. Fayet, *Phys. Lett.* B **89**, 191 (1980).
11. P. Ramond, *Phys. Rev.* D **3**, 2415 (1971); A. Neveu and J. Schwarz, *Nucl. Phys.* B **31**, 86 (1971).
12. J.-L. Gervais and B. Sakita, *Nucl. Phys.* B **34**, 632 (1971).
13. J. Iliopoulos and B. Zumino, B **76**, 310 (1974).
14. P. Fayet and J. Iliopoulos, *Phys. Lett.* B **51**, 461 (1974).
15. P. Fayet, *Phys. Lett.* B **58**, 67 (1975); L. O'Raifeartaigh, *Nucl. Phys.* B **96**, 331 (1975).
16. E. Cremmer et al., *Phys. Lett.* B **147**, 105 (1979).
17. S. Ferrara, D.Z. Freedman and P. van Nieuwenhuizen, *Phys. Rev.* D **13**, 3214 (1976); S. Deser and B. Zumino, *Phys. Lett.* B **62**, 335 (1976).
18. P. Fayet, *Phys. Lett.* B **175**, 471 (1986).
19. P. Fayet, *Nucl. Phys.* B **78**, 14 (1974).

20. M. Gell-Mann, P. Ramond and R. Slansky, *Rev. Mod. Phys.* **50**, 721 (1978).

21. P. Fayet, *Phys. Lett.* B **78**, 417 (1978).

22. P. Fayet, in *History of original ideas and basic discoveries in Particle Physics*, eds. H. Newman and T. Ypsilantis, Proc. Erice Conf., *NATO Series* B **352**, 639 (Plenum, N.Y., 1996).

23. P. Fayet, *Phys. Lett.* B **84**, 416 (1979).

24. L. Girardello and M.T. Grisaru, *Nucl. Phys.* B **194**, 65 (1982).

SUPERSYMMETRIC HIGGS POTENTIALS

LOCHLAINN O'RAIFEARTAIGH

School of Theoretical Physics, Dublin Institute for Advanced Studies,
10 Burlington Road, Dublin 4, Ireland

My first encounter with supersymmetry was at the Warsaw conference on mathematics and physics in March 1974, the forerunner of the present-day M∩P conferences. The initial preprint of Wess and Zumino[1] had just appeared and was a subject of informal discussion. Arthur Jaffe, who immediately saw the potential of the idea, and was organizing the field theory section of the 1974 Aspen summer school, invited me to come to the school and lecture on the subject, on the assumption that by June I would be familiar with it. In this way I was precipitated into learning supersymmetry—and fast. In fact, although I immediately embarked on a study of supersymmetry from the few papers available, there was still much to learn when I arrived in Aspen. There I was overawed to find that not only was there a large and distinguished audience of axiomatic field theorists, but also large and distinguished audiences of string theorists and particle physicists eager to learn about this exotic new subject. I had to work extremely hard to prepare the lectures and I recall staying up an entire night at the Institute in preparation for one particular lecture.

It was evident from the beginning that supersymmetry would be phenomenologically acceptable only if it were broken, but an important question that came up at Aspen was whether it should, or could, be spontaneously broken. The initial attraction of a spontaneously broken theory was that it would allow the interpretation of the neutrino as a Goldstone field and, although this possibility was later ruled out,[2] spontaneous supersymmetry breaking still remained an attractive proposition. At the time the only known mechanism for such a breakdown was the Fayet-Iliopoulos[3] mechanism, and this depended on the presence of an Abelian gauge field. There was no analogue of a Higgs potential i.e. a supersymmetry breaking potential $V(\Phi)$ that depended on (chiral) scalar superfields Φ alone.

In the two months after Aspen I was busy writing up my lecture notes and did not give the supersymmetric Higgs potential much further thought. However, at a conference in Austria that September, during a short review of supersymmetry, I had occasion to discuss the breakdown of internal gauge symmetry in the presence of supersymmetry. To put that discussion into context I should say that, in an elegant 1968 paper called *The Geometry of the Octet*, Michel and Radicati[4] had shown that, for a field ϕ in the adjoint representation, SU(3) broke naturally to U(2) through an equation of the form

$$m\phi + g\phi \wedge \phi = 0 \quad \text{where} \quad (\phi \wedge \phi)_a \equiv d_{abc}\phi^b\phi^c$$

m and g being constant parameters and d_{abc} the Gell-Mann D-matrices. It was obvious that the same applies to the breaking of SU(N) to U(N-1) and the point I was making was that supersymmetry produces the Michel-Radicati equation in a natural way. This happens because the only F-type renormalizable supersymmetric potential for chiral scalar superfields is F^*F, where F is bilinear, and for the adjoint of SU(N) the only bilinear is $F = m\Phi + g\Phi \wedge \Phi$. The snag is that the vacuum is degenerate in the sense that the broken and unbroken solutions, $m\Phi = -g\Phi \wedge \Phi$ and $\Phi = 0$ respectively, both occur at the absolute minima of the potential. Thus, while a spontaneous symmetry breakdown can occur, it is not mandatory. Further, it was possible to show that, to the two-loop approximation at least, the effective potential had the same absolute minima as the original one, and the initial non-renormalization theorems suggested that the degeneracy persisted to all orders in perturbation.

After the lectures I was talking to Abraham Pais, who remarked that, although this phenomenon was interesting and probably merited a letter, the real problem was to find a Higgs potential for supersymmetry itself. For the record, a letter duly appeared in Physics Letters,[5] but by that time Pais' remark had re-awakened my interest in a supersymmetric Higgs potential, and from September onward I began to concentrate on this problem.

There was some conflicting evidence. On the one hand the existence of the Fayet-Iliopoulos mechanism showed that supersymmetry could be spontaneously broken in principle. On the other hand there was a rumored statement by Zumino that there could be no supersymmetry-breaking Higgs potential for a single chiral scalar superfield and, although I had not seen a proof of this statement, I soon convinced myself that it was true. Did this mean that the presence of a gauge field was necessary for supersymmetry breaking? In those early days anything was possible.

The first step forward came with the discovery that with more than one chiral scalar superfield it was possible to construct a potential with degenerate supersymmetry-breaking, that is, a potential whose absolute minimum had both a supersymmetry-preserving and supersymmetry-breaking phase. This was analogous to the Michel-Radicati situation except that the broken and unbroken phases were continuously connected. In fact the potential minimum lay along straight lines and planes (nowadays called flat directions) through the origin in the Φ-plain, with the unbroken minimum at the origin. In spite of the non-renormalization theorems[6] for perturbative quantum corrections, I was convinced that the effective potential would choose the unbroken phase. A supersymmetric version of Murphy's law would surely apply! This belief was reinforced by a letter from Fayet and Sibold, who had seen the preprint of the model, and were of the same opinion.

The failure of the model to produce a mandatory spontaneous breakdown suggested going in the opposite direction and looking for a no-go theorem showing that a supersymmetric Higgs potential would always be degenerate. This direction looked promising but somehow the no-go proofs could never be completed. However, the effort turned out to be worthwhile, because, after one particularly promising attempt, I realized that the flaw in the argument pointed the way toward a counter-example. More precisely, it pointed the way toward a three-field potential of the form $V = F^*F$ with a minimum at $F \neq 0$, a clear signal for a spontaneous breakdown of supersymmetry. In terms of the standard superpotential

$$W = \sqrt{\frac{1}{2}}\lambda_a\Phi_a + \frac{1}{2}m_{ab}\Phi_a\Phi_b + \frac{1}{3}\sqrt{\frac{1}{2}}\Phi_a\Phi_b\Phi_c + \text{herm. conj.}$$

for any number of chiral scalar superfields Φ_a, the parameters chosen were

$$g_{abc} = (g_{ab}\lambda_c + g_{ca}\lambda_b + g_{bc}\lambda_a)/\lambda^2$$

where

$$\lambda_a = \begin{pmatrix} 0 \\ 0 \\ \lambda \end{pmatrix} \qquad g_{ab} = \begin{pmatrix} g & 0 & 0 \\ 0 & 0 & 0 \\ 0 & 0 & 0 \end{pmatrix} \quad \text{and} \quad m_{ab} = \begin{pmatrix} \mu & m & 0 \\ m & 0 & 0 \\ 0 & 0 & 0 \end{pmatrix}$$

After the elimination of dummy fields, the part containing the ordinary chiral scalar fields ϕ_a in the Grassmann variable expansion $\Phi_a = \phi_a + \theta \cdot \psi_a + \dots$ of the superfields then became a potential of the form $2V = F_a^* F_a$, where

$$F_1 = m\phi_2 + \mu\phi_1 + 2g\phi_3\phi_1 \qquad F_2 = m\phi_1 \qquad F_3 = \lambda + g\phi_1^2$$

and led to a potential of the form

$$2V = \lambda^2 + (m^2 + 2\lambda g)(\text{Re}\phi_1)^2 + (m^2 - 2\lambda g)(\text{Im}\phi_1)^2$$
$$+ g^2(\phi_1^*\phi_1)^2 + F^* F_1$$

This potential did indeed produce a spontaneous breakdown of supersymmetry.[7] Later, of course, more sophisticated models based on the same principle were constructed.

Thus the sequence of discovery of a supersymmetric Higgs potential was from first trying for positive result for more than one superfield, then trying for a negative result for any number of superfields, and ending up with a positive result for three superfields.

The story does not quite end at this point, because (2) is only the classical potential. To copperfasten the result it had to be shown that the potential was the most general renormalizable potential consistent with all the available symmetries, since otherwise quantum corrections could induce new terms and possibly restore the supersymmetry. What guaranteed the generality of the potential was the existence of R-symmetry (phase-symmetry of the Grassmann variables). In fact the above potential is the most general one of its kind consistent with supersymmetry and R-symmetry. Furthermore, it could be shown that, at the perturbative level at least, the R-symmetry was not broken by the quantum corrections.

In fact the perturbative corrections and the general question of renormalization was soon afterwards investigated in a systematic and clear manner by Piguet, Sibold and their collaborators.[8,9] A decade later it was shown by Affleck et al.[10] that the R-symmetry and the mechanism as a whole were preserved by a class of non-perturbative quantum corrections.

Both the potential V above and the Fayet-Iliopoulos potential break the supersymmetry at tree-level i.e. they break it at the level of the classical action, or, more precisely, at the level of the action in the path-integral. This raises the question as to whether, for a classically unbroken supersymmetry, a spontaneous breakdown could be induced by the quantum corrections. The usual non-renormalization theorems indicate that this does not happen perturbatively (there is no analogue of the Coleman-Weinberg mechanism) but it might happen at the non-perturbative level. In fact it has been shown that, in the presence of gauge fields, a spontaneous breakdown of supersymmetry can be induced in principle by instantons and anomalies.[6,10]

Whether some modern version of either the tree-level or the non-perturbative models have anything to do with reality is, of course, another question!

References

1. J. Wess and B. Zumino, *Phys. Lett.* B **49**, 39 (1974).
2. B. deWit and D. Freedman, *Phys. Rev. Lett.* **35**, 827 (1975).
3. P. Fayet and J. Iliopoulos, *Phys. Lett.* B **51**, 461 (1974).
4. L. Michel and L. Radicati, *Ann. Phys.* **66**, 758 (1971).
5. L. O'Raifeartaigh, *Phys. Lett.* B **56**, 41 (1975).
6. S. Weinberg, *The Quantum Theory of Fields III*, Cambridge Univ. Press, 2000.
7. L. O'Raifeartaigh, *Nucl. Phys.* B **96**, 331 (1975).
8. T. Clarke, O. Piguet, and K. Sibold, *Nucl. Phys.* B **119**, 292 (1977).
9. O. Piguet, K. Sibold, and M. Schweda, *Nucl. Phys.* B **168**, 337 (1980).
10. I. Affleck, M. Dine, and N. Seiberg, *Nucl. Phys.* B **256**, 557 (1985).

THE SUPERSYMMETRIC EFFECTIVE POTENTIAL

PETER WEST

King's College, Department of Mathematics, University of London,
Strand, London WC2R 2LS, United Kingdom

I would first like to thank Misha Shifman for his kind invitation to write an article on the paper of Ref. 1 which had the above title. This paper proved that if supersymmetry is preserved at the classical level then the effective potential in a supersymmetric theory had no perturbative quantum corrections. As the paper pointed out, this meant that the degeneracies that were generic in the classical potentials of supersymmetric theories were not removed by perturbative quantum corrections; in modern language supersymmetric theories had moduli. This result, together with earlier work, provided the theoretical basis for the realization that supersymmetry solved the technical hierarchy problem.[2] In other words, the scalar masses in supersymmetric grand unified theories did not receive large corrections and as a result in supersymmetric theories, once one set a small scale for electro-weak breaking, quantum corrections did not lead to corrections of order the grand unified scale. It also meant that supersymmetry could not be broken by perturbative quantum corrections if it was not broken at the classical level and as a result it placed considerable constraints on the way realistic supersymmetric models could be constructed.

I would like to concentrate on explaining the developments that lead up to this paper and at the same time try to give some feeling for what it was like to be a a Ph.D. student during the early days of supersymmetry. After I had finished my undergraduate studies at Imperial College in the summer of 1973, I had the good fortune to be accepted as a Ph.D. student of Abdus Salam who was a member of the theoretical physics group of Imperial College and also the director of the International Center for Theoretical Physics in Trieste. Abdus Salam spent most of his time in Trieste, but he did return to London each month. As was normal for students who had not done Part III in Cambridge, I spent most of my time during my first year taking courses at Imperial and King's Colleges. Even in those days there was a large gap between what was in the undergraduate program and the material that was required to do research. Clearly, there was no material on supersymmetry, but there was also no course on string theory, since this was not viewed as an important subject and perhaps more surprisingly no lectures on the spontaneous symmetry breakdown and the developments we now call the Standard Model. However, we did have lectures on quantum field theory, QED, general relativity, group theory and hadronic dual models and Regge poles. The course on field theory was taught by Tom Kibble and among many other topics he explained the problem of infinities in field theories and the renormalization program. I was somewhat surprised to find that the most promising theories of nature required such unesthetic manipulations. One of the hot topics at that time was the discovery of a new resonance which was eventually identified as the bound state of the charm quark, but at Imperial, which had perhaps the best theory group in England, a debate raged as to whether it might be the fourth color.

Despite the somewhat reduced content compared to today, I was fully occupied studying these courses and so there was little reason to see Abdus Salam. However, once the courses had finished in the summer of 1974, I went to see him to find a suitable topic to work on. Although supersymmetry had been discovered in 1972 by Golfand and Likhtman[3] and independently by those working in string theory,[5] there was very little development, except for the paper of Volkov and Akulov,[4] until the paper of Wess and Zumino in 1974.[6] This latter paper found a four dimensional quantum field theory with this symmetry which became known as the Wess-Zumino model. It was with this advance that small groups of physicists, mainly in France, Germany and England, began to explore the consequences of supersymmetry in earnest. An early result was the discovery of a supersymmetric theory that included a vector particle, the $\mathcal{N} = 1$ Yang-Mills theory.[7] Just before I went to see Abdus Salam, he and John Strathdee had written a paper[8] introducing superspace. This provided a manifestly covariant description of the Wess-Zumino model and later all supersymmetric theories. At the end of this paper was an appendix giving the Feynman rules for this theory in the superspace formalism. These became known as the super-Feynman rules. Abdus Salam suggested I should study this paper. Given my preference to study a subject that did not possess such undesirable features as infinities, I had an initial reluctance. However, once I started reading the paper I became fascinated with these new models and the superspace formalism.

Even today, with the benefit of a number of pedagogical text books, learning supersymmetry is not easy. It involves a systematic level of manipulation of Clifford algebras, such as Fierz rearrangements, that is substantially more involved than previous developments. Hence studying the paper of Strathdee and Salam was not straightforward. However, being at Imperial had another very substantial advantage; Bob Delbourgo was a member of the staff. He explained a number of the technicalities and provided me with some sheets, that contained all the required identities. In effect, he become my supervisor while Abdus Salam was not there. As a result, I was able to start rederiving some of the results in the Salam and Strathdee paper and begin doing research.

Initially, Bob thought he might have a problem I could work on. Just before then it had been shown that the Wess-Zumino model possessed only one infinity when, even if supersymmetry was preserved, one might have expected that it should have three infinities. This result was first shown at one loop by Iliopoulos, Wess and Zumino[9] and to all orders by Iliopoulos and Zumino.[10] The idea of Bob was to use the new super-Feynman rules of Salam and Strathdee to rederive this result. Understandably, he found the problem too tempting and he solved it within a few days.[11] The result was the first paper using the new technique of super-Feynman rules.

Hence at this stage I was still to find a specific problem to work on. Since in supersymmetric theories the masses of all the particles in the same supermultiplet were equal, one of the early problems was how to spontaneously break supersymmetry. It was clear that the potentials in supersymmetric theories were much more constrained than those of a generic quantum field theory. For example, they were positive definite. While breaking internal symmetries was not so problematic, it was far from clear how to break supersymmetry. However, it was found that by a careful choice of field content and interactions one could break supersymmetry spontaneously at the tree level. This was achieved in two different ways; one was discovered by Fayet and Iliopoulos[12] and the other was due separately to Fayet and O'Raifeartaigh.[13] Although the masses of the particles that emerged from these breakings were not the same in each supermultiplet, they did obey a pattern of masses that was rather constrained and not favorable from a model building viewpoint. It had been thought that things would be better in supersymmetric models in which supersymmetry was preserved at the classical level, but spontaneously broken at the quantum level.

We are so familiar with properties of supersymmetric theories today that it is worth recalling the thinking at that time. Most of the quantum field theories that had been studied before the advent of supersymmetry had potentials which uniquely fixed the vacuum of the theory at the classical level up to the symmetries of the theory. The effective potential of quantum field theories received quantum corrections at all orders of perturbation theory and the one loop corrections had a particularly elegant form due to Coleman-Weinberg.[14] To calculate these latter effects one first found a minimum of the classical potential and then computed the masses of the particle in terms of the shifted scalar fields from the minimum. In some exceptional cases, the potentials might have degenerate minimum at the classical level, but these would be removed by quantum corrections.

One very usual feature of the potentials in many supersymmetric theories was that the classical potentials did not uniquely determine the vacua of the theory even up to the symmetries of the theories. This could be viewed as a consequence of the very constrained nature of the superpotentials which meant that they often had a larger symmetry than the theory as a whole. The potentials were said to have flat directions, that is directions along which the value of the potential was a minimum. This was viewed as an undesirable feature and another reason to calculate the effective potential in supersymmetric theories was to see how these degeneracies were removed by the quantum corrections.

Abdus Salam and John Strathdee suggested that Bob Delbourgo and I should compute the one loop effective potential in some supersymmetric models in which supersymmetry was not broken at the classical level to see if it was broken by quantum corrections. They also provided us with a number of promising models. To do this we had to compute the Coleman-Weinberg effective potential as described above. We discovered that the one loop effective potential had a habit of vanishing. To find a model that "worked" we were forced to consider more complicated models with many fields. To compute the effective potential became more and more complicated requiring the diagonalization of large matrices one of which was 26 by 26 dimensional. We displayed these matrices on large sheets of drawing paper and patiently swapped rows and columns until it was diagonal. After taking more than a week to diagonalize such a matrix we found that the result was a complex effective potential. This meant that we had begun from the wrong minimum. It was only after this defeat that it became clear that one had to think about the problem in a more fundamental way.

It was at this point that I began to wonder if the effective potential in a supersymmetric theory might actually vanish. By writing out the part of the Feynman graph that involved the Grassmann odd components of superspace it became clear that this was the case. Initially, I did this loop by loop, but eventually the proof of the all orders result became apparent.[1]

Whenever I learnt that Abdus Salam was back on one of his visits to Imperial College, I would knock on his door. He was almost always available and keen to discuss physics in a very friendly way. Since I had not really made much significant progress in my research up until then, I was rather relieved to have found a result which I thought was of some significance. Somewhat to my surprise Abdus Salam did not share my enthusiasm. However, on his return to Trieste he explained the result to John Strathdee who doubted it was true. After some checks they did agree it was correct.

At this point I still had six months left of the three years which it usually took to get a Ph.D. I began working with Ali Chamseddine who was also a Ph.D student at Imperial. Supergravity had just been discovered[15] and we decided to try to construct supergravity by gauging the super Poincare group. Although we were aware of the paper of Tom Kibble which gave the first order formalism of general relativity in which the spin-connection was the gauge field for the Lorentz group, we were too naive to realize that gauging of space-time groups, such as the Poincare group, to gain a complete theory of gravity had not been attempted. We soon found[16] that, with one assumption, gauging the super Poincare group did lead to the theory of supergravity already proposed. This method had one other advantage, it gave the first algebraic proof of the invariance of supergravity.

After my Ph.D had finished in the summer of 1976 I had to find a job, which in those days were few and far between. I tried to get a research council fellowship, but in the interview I was asked about the detailed connection of supersymmetry to the known particles. I was not successful since I, and indeed no one else, had much idea of this connection. This attitude was in keeping with the general belief in that part of the research council responsible for theoretical physics that only work rather directly related to experiment should be funded. Fortunately, I was given a Royal Society postdoctoral fellowship to go to the Ecole Normale in Paris for a year and so was able to continue working in supersymmetry.

The following does not constitute a complete list of references, but with the exception of Ref. 2, contains those I found most useful during my Ph.D.

References

1. P. West, *Nucl. Phys.* B **106**, 219 (1976).
2. E. Witten, *Nucl. Phys.* B **186**, 513 (1981), S. Dimopoulos and H. Georgi, *Nucl. Phys.* B **193**, 150 (1982), N. Sakai and Z. Physik, C11 (1982) 153, R. Kaul, *Phys. Lett.* B **109**, 19 (1982).
3. Y.A. Golfand and E.S. Likhtman, *JETP Lett.* **13**, 323 (1971).
4. D.V. Volkov and V.P. Akulov, *Pis'ma Zh. Eksp. Teor. Fiz.* **16**, 621 (1972); *Phys. Lett.* B **46**, 109 (1973).
5. P. Ramond, *Phys. Rev.* D **3**, 2415 (1971), 2415 (1971); A. Neveu and J.H. Schwarz, *Nucl. Phys.* B **31**, 86 (1971), 86 (1971); *Phys. Rev.* D **4**, 1109 (1971); J.-L. Gervais and B. Sakita, *Nucl. Phys.* B **34**, 477, 632 (1971); F. Gliozzi, J. Scherk, and D.I. Olive, *Nucl. Phys.* B **122**, 253 (1977).

6. J. Wess and B. Zumino, *Nucl. Phys.* B **70**, 139 (1974).
7. S. Ferrara and B. Zumino, *Nucl. Phys.* B **79**, 413 (1974); A. Salam and J. Strathdee, *Phys. Rev.* D **11**, 1521 (1975).
8. A. Salam and J. Strathdee, *Phys. Rev.* D **11**, 1521 (1975); *Nucl. Phys.* B **86**, 142 (1975).
9. J. Wess and B. Zumino, *Phys. Lett.* B **49**, 52 (1974).
10. J. Iliopoulos and B. Zumino, *Nucl. Phys.* B **76**, 310 (1974).
11. R. Delbourgo, *Nuovo Cim.* **25**, 646 (1975). This result was also given in D. M. Capper, *Nuovo Cim.* **25**, 259 (1975).
12. P. Fayet and J. Iliopoulos, *Phys. Lett.* B **51**, 461 (1974).
13. L. O'Raifeartaigh, *Nucl. Phys.* B **96**, 331 (1975). P. Fayet, *Phys. Lett.* B **58**, 67 (1975).
14. S. Coleman and E. Weinberg, *Phys. Rev.* D **7**, 1888 (1973).
15. D. Freedman, P. van Nieuwenhuizen, and S. Ferrara, *Phys. Rev.* D **13**, 3214 (1976); *Phys. Rev.* D **14**, 912 (1976); S. Deser and B. Zumino, *Phys. Lett.* B **62**, 335 (1976).
16. A. Chamseddine and P. West, *Nucl. Phys.* B **129**, 39 (1977).

5

The Historical Perspective

M. Marinov

R. Di Stefano

REVEALING THE PATH TO THE SUPERWORLD[a]

M. S. MARINOV

Physics Department, Technion – Israel Institute of Technology
Haifa, 32000 Israel

A brief history of ideas which paved the way to supersymmetry is presented.

One of the most impressive discoveries of the 20th century is the understanding that neither real nor complex numbers are sufficient to describe physical systems. It was found that, besides the usual coordinates, electrons and other elementary particles have spin degrees of freedom, which have no intuitive classical analogues, and *anti-commuting variables* must be employed to produce a proper and elegant representation of dynamics at the microscopical level. *The Superworld* has commuting and anti-commuting dimensions, owing to (or giving rise to) two kinds of particles, bosons and fermions.

The way to this discovery was paved by the mathematics of the 19th century. In 1843 William R. Hamilton discovered quaternions, an associative algebra of number quadruples with three *anticommuting* imaginary units, including division (each nonzero element has its inverse). The discovery was accepted as a miraculous revelation — "the light dawned upon Hamilton on a certain October day in Dublin," as some of his adepts recalled.[1] The complete theory, including elements of the vector analysis, was published in *Lectures on Quaternions* (1853) and in a more exhaustive way in *Elements of Quaternions* (1866).[2] About three decades after the work on quaternions was first published, William K. Clifford introduced a more general associative algebra (though with no division in general, unlike that of quaternions) of *bi-quaternions* with n auxiliary anticommuting units. The bi-quaternions were presented in the form

---

[a]This is a revised version of the article prepared by M. Marinov for the Yuri Golfand Memorial Volume *The Many Faces of the Superworld*, (World Scientific, Singapore, 2000), page 32. M. Marinov worked on it in the summer of 1999, and then started revising it to include a part devoted to supergravity. The work was not completed: he died of cancer on January 17, 2000.

$c = a + \sum_{k=1}^{n} b_k \epsilon_k$, where a and b_k are real and ϵ_k^2 may be $+1, -1$ or 0. The latter is the case used in functional formalism for Fermi-type variables and in the supersymmetry in particular. In general, the Clifford algebra is given by the anti-commutation relations for its generators, $\epsilon_k \epsilon_l + \epsilon_l \epsilon_l = g_{kl}$, where the quadratic form given by g_{kl} may be degenerate. The matrices introduced in physics by Pauli and Dirac are two examples of the Clifford algebras with non-degenerate g_{kl}.

The introduction of quaternions by the Royal Astronomer of Ireland was a spectacular achievement, but at about the same time a less celebrated mathematician, Hermann G. Grassmann, a high school teacher in Stettin, Prussia, developed a more general, profound and abstract "theory of extension." Grassmann was an outstanding person of encyclopedic knowledge in science, mathematics and philology (one of his famous works was a dictionary of Sanskrit)[3]. The treatise describing Grassmann's theory was published[4] in 1844, and the most complete second edition appeared in 1878. Both Hamilton and Grassmann discovered vector analysis, obtained new associative algebras, and found a number of other beautiful things. A struggle for priority (not the first one in the history of mathematics) between the British and German mathematical communities continued for decades. In fact, the "theory of extension" provided a basis for modern differential geometry and tensor analysis. It was very much ahead of its time, but now it is studied in some form at most departments of mathematics. Referring to the Clifford algebra with $\epsilon_k^2 = 0$, Berezin introduced the term "Grassmann algebra with anti-commuting generators," which is widely used now in theoretical physics.

Until the beginning of the 20th century Grassmann's theory was available to only a few initiated persons, since Grassmann's book was hardly easy reading. However, we have evidence that a great physicist studied from it as a boy; that was 16-year-old Enrico Fermi. Emilio Segré published a letter by Adolfo Amidei, a colleague and friend of Enrico's father, in his scientific biography of Fermi[5]. We read in the letter:

I also deemed it appropriate for him [Enrico] to study the *Ausdehnungslehre* by H. Grassmann, which has an introduction on the operations to deductive logic by Giuseppe Peano. These books were lent to him in 1918. I thought it appropriate because it was my opinion that the *Ausdehnungslehre* [similar to vector analysis] was the most suitable tool for the study of different branches of geometry and theoretical mechanics, because in this system of calculus the operations are not performed on numbers [coordinates] which define the geometrical objects but on the objects themselves, thus obtaining formulae which are simple and of easy interpretation. Furthermore, this calculus contains, as a particular case, analytical geometry, and is specially suited to the study of theoretical mechanics and graphical statics... Enrico found vector analysis very interesting, useful and not difficult...

Now one may only speculate if Fermi had in mind Grassmann's construction when he elaborated the theory of degenerate "fermion gas".[6] No reference to anti-commutation relations was given. It was in Göttingen that the canonical anti-commutation relations (CAR) were consistently introduced by Pascual Jordan and Eugene Wigner[7] as a necessary addition to the canonical commutation relations (CCR) of the Heisenberg–Weyl algebra. Thus, commuting and anti-commuting canonical operators were put on an equal footing. Boson and fermion fields appeared in conventional quantum field theory.

The operator formalism, including the standard Fock space in field theory, is based consistently upon CCR and CAR, but in order to employ the functional formalism of Schwinger and Feynman one needs a further development of the corresponding classical theory. In the boson sector, where CCR correspond to the standard Poisson brackets, the classical limit leads to commuting quantities, so the operators (q-numbers) are reduced to complex numbers (c-numbers), where the standard calculus is known. In the fermion sector, the classical limit leads to "anti-commuting c-numbers," while the Clifford algebra with $\epsilon_k^2 = +1$ is reduced to the degenerate case where $\epsilon_k^2 = 0$ (since CAR are proportional to $\hbar$), and the necessary elements of calculus need an extension. First, Julian Schwinger advocated the universality of his "quantum action principle" in a reply[8] to opponents who doubted its applicability to fermion fields. According to Schwinger's arguments, the canonical variables may be of two kinds; first-kind variables with anti-symmetrical Poisson brackets leading to CCR upon quantization, and second-kind variables with symmetrical Poisson brackets leading to CAR. The general principle may be formulated in the following way. The canonical variables of the boson type x_a (the usual phase-space vector) and the fermion type (anti-commuting) ξ_α are united in one vector $X_A \equiv (x, \xi)$, where the index is $A \equiv (a, \alpha)$, of the extended phase space (which may be called "phase superspace"). For any pair of the dynamical functions $\mathcal{A}(X)$ and $\mathcal{B}(X)$, the Poisson brackets were defined

$$\{\mathcal{A}, \mathcal{B}\}_{\text{P.B.}} \equiv \frac{\partial \mathcal{A}}{\partial X_A} \frac{\partial \mathcal{B}}{\partial X_A} \Omega_{AB}. \tag{1}$$

The (X-independent) bilinear form Ω must be nondegenerate and skew-Hermitian, i.e. $\Omega^\dagger = -\Omega$. Its canonical real form is $\Omega_{ab} = \omega_{ab} = -\omega_{ba}$, $\Omega_{\alpha\beta} = i\delta_{\alpha\beta} = i\delta_{\beta\alpha}$, $\Omega_{a\alpha} = 0 = \Omega_{\alpha a}$. The Poisson brackets defined in this way constitute a Z_2-graded Lie algebra of the simplest type, provided the partial derivatives in ξ are defined properly. In agreement with Grassmann's concepts, Schwinger considered the second-kind variables as elements of the *exterior algebra* (another name for Grassmann's algebra of differential forms). The formalism was developed in a series of papers published in *Proc. Natl. Acad. Sci.*, which were later reprinted and commented on in Schwinger's book.[9] In these terms, the formulation of a linear realization of supersymmetry looked straightforward. Let us assume that the system is described by a quadratic Hamiltonian $\mathcal{H}(X) = \frac{1}{2}(X, \breve{H}X)$, where $\breve{H}$ is a linear operator in the extended phase space. The symmetry of the system under linear transformations $X = \breve{L}Y$ is given by the Poisson-bracket-generated Lie algebra of *all* quadratic functions $\mathcal{C}(X) = \frac{1}{2}(X, \breve{C}X)$ which are in involution with $\mathcal{H}(X)$, namely

$$\{\mathcal{H}, \mathcal{C}\}_{\text{P.B.}} = 0. \tag{2}$$

Evidently, besides the blocks leaving the boson and fermion sectors invariant, the generic operator $\breve{C}$ satisfying the above condition contains a part mixing x and ξ and proportional to anti-commuting parameters. Actually, all the quadratic functions $\mathcal{C}(X)$, which are the integrals of motion for the system of harmonic oscillators of the boson and fermion type, constitute a simple Z_2-graded Lie algebra with respect to the Poisson brackets. Evidently, the foundation of supersymmetry was elaborated by Schwinger, but further development came from another side.

Feynman's functional integral in QED was restricted to electromagnetic (boson) field A_μ. For instance, integrating out A_μ, which appears quadratically in the Lagrangian, one gets an effective (nonlocal) action functional for the electron (fermion) field ψ. The original QED Lagrangian was bilinear also in ψ, so why not start from the integral in ψ? This was the question asked by Isaak Khalatnikov, and he found[10] that it may be done, indeed — the result would be consistent with the canonical operator formalism, provided that the Gaussian integral in the anti-commuting variables ψ results in the determinant in the numerator, and not in the denominator, as in the usual calculus. Khalatnikov presented his work at the seminar steered by Landau, which was attended by Felix Berezin, then a student of mathematics under the supervision of Israel Gelfand. The job performed by Berezin was to make Khalatnikov's conjecture consistent mathematically. The arguments were quite simple. A way to evaluate the usual Gaussian integral

$$I(B) \equiv \int_{\mathsf{R}_n} \exp[-(x, Bx)]d^n x \qquad (3)$$

in the n-dimensional Euclidean space R_n with the standard measure $d^n x$ and a positive definite linear operator $B = \tilde{B}$ is to employ its invariance. After any linear change of variables $x = Ly$, where $\det L \neq 0$, one gets the integral of the same type, so

$$I(B) = I(\tilde{L}BL)\det L, \quad \forall L, \qquad (4)$$

where the second factor is due to the Jacobian, as $d^n x = d^n y \det L$. Hence, one concludes that

$$I(B) = [\det B]^{-1/2} \times \text{const.} \qquad (5)$$

Here the constant is known, of course, but it is practically irrelevant for the functional calculations. Now let us consider the integral in the anti-commuting variables and *define* it in such a way as to get the desired result,

$$J(F) \equiv \int_{\mathsf{G}_n} \exp[-(\xi, F\xi)]d^n\xi = [\det F]^{1/2} \times \text{const}, \qquad (6)$$

where $F = -\tilde{F}$ is a linear operator in the anti-commuting space G_n, and the integration measure should be defined properly. Apparently, in order to get the result one has to *assume* that under the linear substitution $\xi = L\eta$ one must have $d^n\xi = d^n\eta/\det L$. Therefore, as soon as for the anti-commuting variables $\prod_{k=1}^n \xi_k = \det L \prod_{k=1}^n \eta_k$, the integral $\int(\prod \xi)d^n\xi$ is an invariant of the linear transformations. Consequently, Berezin *postulated* the basic properties of the the elementary integrals,

$$\int \xi d\xi = 1, \quad \int d\xi = 0. \qquad (7)$$

These consistent rules opened the way to complete calculus in the anti-commuting variables, since the partial derivatives in ξ were defined straightforwardly, and in applications to quantum field theory. The results appeared in a series of Berezin's papers and in the book[11] published in Russian in 1965, which was translated immediately. A work on the extension of the Lie groups to the groups with commuting and anti-commuting parameters (now called the super-Lie groups) was also published by Berezin and Katz.[12] The ground for supersymmetry was prepared; further progress was prompted by a new concept.

About 1969, the dual reggeon–resonance model of strong interactions, found by Veneziano and developed by a number of people, was formulated in terms of four-dimensional harmonic oscillators. In constructing the Fock space, the problem was to decouple some ghost states of the negative norm, owing to the time-like components of the oscillators. Constraints had to be imposed on the physical states. As was shown by Virasoro, the consistency of the approach resulted from the structure of an infinite-dimensional Lie algebra. After Nambu advanced the idea that the oscillators represent relativistic string, the scheme was reformulated as a field theory in the (τ, σ) space, where the field was the Minkowski coordinates $x_\mu(\tau, \sigma)$

of the points on the world sheet of the string. Respectively, the Virasoro algebra was found to be the Lie algebra of the group of the conformal transformations on the string world sheet. In QED the gauge symmetry is known to be necessary for consistent quantization of the four-vector potential, and therefore the conformal symmetry was also called the *gauge symmetry* in the dual models. In order to include fermions and pion-like negative-parity bosons, Neveu, Schwarz[13] and Ramond[14] introduced fermion oscillators besides the boson oscillators, and the problem of decoupling of the ghost states arose again. Neveu, Schwarz and Thorn[15] showed that the proper Fock space is constructed by means of anti-commuting operators $\hat{G}_r$ satisfying the following algebra, together with the previously known Virasoro operators $\hat{L}_n$:

$$\left[\hat{G}_r, \hat{G}_s\right]_+ = -2\hat{L}_{r+s} + a_1\hat{I}\delta(r+s), \qquad (8)$$

$$\left[\hat{G}_r, \hat{L}_m\right]_- = (r - m/2)\hat{G}_{r+m},$$

$$\left[\hat{L}_m, \hat{L}_n\right]_- = (m - n)\hat{L}_{n+m} + a_0\hat{I}\delta(n+m).$$

Here (m, n) are integers, (r, s) are both integers or half-integers, $\hat{I}$ is the unit operator, and $\delta(k) = 1$ for $k = 0$ and 0 for $k \neq 0$. The additional terms with constants a_0 and a_1 (their values are irrelevant now) were discovered later. They are due to quantum anomalies and do not appear in the classical Poisson brackets. In Ref. 15 the operators $\hat{L}_m$ and $\hat{G}_r$ were represented as bilinear combinations of the oscillator creation–annihilation operators, so that Eqs. (8) result from CCR and CAR. In order to account for the fermion degrees of freedom in the framework of string theory, the anti-commuting fields $\psi_\mu(\tau, \sigma)$ were introduced to represent the fermion oscillators. As in the purely bosonic case, the symmetry providing consistent constraints decoupling the ghost states had a geometrical meaning discovered by Gervais and Sakita.[16] In an appropriate (conformal) gauge, it was just an infinite-dimensional group with commuting and anti-commuting parameters acting *linearly* upon x_μ and ψ_μ. Thus, the Z_2-graded Lie algebra (8) was interpreted by Gervais and Sakita as the algebra of generators of the field transformations. This property of the boson and fermion fields on the string world sheet was called by the authors the *supergauge* symmetry, which was probably the first application of the word "super" in this context. It is noteworthy that Gervais and Sakita referred to the textbook[17] by Jan Rzewuski, where Berezin's functional formalism was described in full detail. Thus, Berezin's ideas influenced the work on supersymmetry.

In two-dimensional space–time the group of conformal (and superconformal) transformations is infinite-dimensional, but it has a very simple realization. It is just the group of diffeomorphisms of the light-cone coordinates. In the four-dimensional Minkowski space the dimensionality of the conformal group is 15, and its geometrical action is more intricate. Given the conformal group, can one define the generators of the superconformal transformations so that their anticommutators would be the generators of the conformal group in four dimensions? This was the question asked by Julius Wess and Bruno Zumino,[18] and, miraculously, they found the positive and complete answer. The four-dimensional superconformal transformations were represented linearly on multiplets having equal numbers of boson and fermion fields. It looked like a miracle that the Lie algebra of the generators was closed and included the conformal generators as a subalgebra. Only later, when the formalism of superspace and superfields was developed, was the miracle explained properly. As soon as the transformations were realized linearly in terms of the local fields, the generators were represented as integrals of the corresponding supercurrents over a space-like surface, like the four-momentum, which is the integral of the energy–momentum tensor. Examples of the invariant Lagrangians were also constructed, and it appeared that the key to the solution was the introduction of auxiliary fields, which enter the theory without derivatives and have trivial equations of motion. That was an unusual and a very important feature, which enabled the linear realization of the symmetry. It was soon found that the Poincaré group, as a subgroup of the conformal group, has also a nontrivial superextension. The use of the geometrical ideas enabled the construction of the beautiful theory of supersymmetry and supergravity, which is presented now in a number of books (e.g. by Wess and Bagger[19]).

Yuri Golfand discovered the super-Poincaré algebra in his own way, at least three years before the wonderful work by Wess and

Zumino was published, and even before the works[15,16] on supersymmetry in dual models were done. It is known that Golfand discussed the new symmetry with his colleagues in the late 1960's, trying to solve the puzzle of weak interactions, before the electroweak theory did the job. That is why the parity violation was incorporated as well in the algebra. Though Golfand lived in Moscow and belonged to the same community as Khalatnikov and Berezin, his approach was quite independent and the arguments were more aesthetical than pragmatical. The known basic fields were those representing vector bosons (like photons) and fermion spinors (like electrons). By the standard gauge invariance, the vector fields are intimately related to the four- momentum operators. Are there operators associated with the fermion fields? If such operators exist, they must be spinorial and satisfy anti- commutation relations, naturally involving the momentum operators in the right-hand side. The crucial question was whether one could construct a super-Lie algebra, so that the anti-commutators would satisfy the Jacobi identities. This was checked by Golfand and Likhtman, and their first note was published.[20] As soon as the algebra was constructed, the next problem was to realize it in terms of the local fields and to construct an invariant theory. This program was formulated and the first supersymmetric model was constructed.[20] The general scheme of supersymmetric model-building was also proposed, in a recursive way,[21] since Golfand and Likhtman had no linear realization for supersymmetry. This was done with auxiliary fields by Wess and Zumino. They started from the local fields and constructed supersymmetric field theory independently. Their work[18] then triggered the subsequent explosive development. Wess and Zumino were inspired by Neveu, Schwarz and Ramond. At the time of their work they knew neither of Golfand nor of any other research on four-dimensional supersymmetry.

In fact, an independent research was carried out in 1972 by D.V. Volkov and V.P. Akulov who posed the question whether Goldstone particles with spin 1/2 could exist. They formulated[22] a nonlinear realization of the super-Poincaré algebra and suggested a Lagrangian for the self-interacting Goldstino field, the Goldstone fermion of (the spontaneously broken) supersymmetry. The idea of presenting neutrino as a Goldstino was then abandoned because of phenomenological difficulties. Thus, unfortunately, this work went relatively unnoticed, producing no impact in the outside world. It was revived only after Wess and Zumino.

Gauging the super-Poincaré group (i.e. making supersymmetry local) leads to supergravity. A description of gravity based on the minimal supersymmetry and including fields of spin 3/2, 1 and 1/2 was suggested as early as in 1973 by Volkov and Soroka.[23] A comprehensive theory of supergravity was worked out by[b] [Freedman *et al.*[24] and by Deser and Zumino[25]].

Almost three decades after its discovery, supersymmetry remains one of the most beautiful theories in physics. Is it just a theoretical option, or a symmetry really existing in Nature? This is probably one of the most important questions in modern particle physics.

Acknowledgments

I am grateful to Misha Shifman for encouragement and helpful advice. Support for the work from the Fund for Promotion of Research at the Technion is acknowledged.

References

1. D. J. Struik, *A Concise History of Mathematics* (Dover, New York, 1948).
2. W.R. Hamilton, *Elements of Quaternions*, 2 vols., (London, 1866); 3rd edition (Chelsea Pub. Co. New York, 1969).
3. *Hermann Günther Grassmann (1809–1877): Visionary Mathematician, Scientist and Neohumanist Scholar*, Ed. G. Schubring (Kluwer, Dordrecht–Boston–London, 1996).
4. H.G. Grassmann, *Die lineale Ausdehnungslehre, ein neuer Zweig der Mathematik* (O. Wigand, Leipzig, 1844); 2nd rev. edition, 1878.
5. E. Segré, *Enrico Fermi — Physicist* (University of Chicago Press, Chicago–London, 1970), p. 10.
6. E. Fermi, *Rend. Lincei* **3**, 145 (1926); *Z. Physik* **36**, 478 (1926).

---

[b]This paragraph was not completed in the manuscript.

7. P. Jordan and E. Wigner, *Z. Physik* **28**, 631 (1928).
8. J. Schwinger, *Phil. Mag.* **44**, 1171 (1953).
9. J. Schwinger, *Quantum Kinematics and Dynamics* (W. A. Benjamin, New York, 1970).
10. I.M. Khalatnikov, *JETP* **28**, 635 (1954).
11. F.A. Berezin, *Method of Second Quantization* (Academic, NY, 1966).
12. F.A. Berezin and G. I Katz, *Mat. Sbornik* **82**, 343 (1970) [*Math. USSR – Sbornik*, **11**, 311 (1970)].
13. A. Neveu and J.H. Schwarz, *Nucl. Phys.* **B31**, 86 (1971).
14. P. Ramond, *Phys. Rev.* **D3**, 2415 (1971).
15. A. Neveu, J.H. Schwarz, and C.B. Thorn, *Phys. Lett.* **35B**, 529 (1971).
16. J.-L. Gervais and B. Sakita, *Nucl. Phys.* **B34**, 632 (1971).
17. J. Rzewuski, *Field Theory II* (Iliffe Books, London, 1969).
18. J. Wess and B. Zumino, *Nucl. Phys.* **B70**, 39 (1974).
19. J. Wess and J. Bagger, *Supersymmetry and Supergravity*, 2nd edition, (Princeton University Press, Princeton, 1992).
20. Yu.A. Golfand and E.P. Likhtman, *JETP Letters* **13**, 323 (1971) [Reprinted in *Supersymmetry*, Ed. S. Ferrara (North-Holland/World Scientific, Amsterdam–Singapore, 1987), Vol. 1, p. 7.]
21. Yu.A. Golfand and E.P. Likhtman, in I.E. Tamm Memorial Volume, *Problems of Theoretical Physics* (Moscow, Nauka, 1972), p. 37 [Engl. translation in *The Many Faces of the Superworld*, Ed. M. Shifman (World Scientific, Singapore, 2000)].
22. D.V. Volkov and V.P. Akulov, *Phys. Lett.* **B46**, 109 (1973).
23. D.V. Volkov and V. Soroka, *JETP Lett.* **18**, 312 (1973).
24. D.Z. Freedman, P. van Nieuwenhuizen, S. Ferrara, *Phys. Rev.* **D13**, 3214 (1976).
25. S. Deser, B. Zumino, *Phys. Lett.* **62B**, 335 (1976).

NOTES ON THE CONCEPTUAL DEVELOPMENT OF SUPERSYMMETRY [a]

ROSANNE DI STEFANO

Department of Physics and Astronomy,
Tufts University
Medford, MA 02155
and
Harvard-Smithsonian Center for Astrophysics
Cambridge, MA 02138

A short history of the conceptual development that led to the notion of supersymmetry is presented, and some of the developments that arose from the growing understanding of supersymmetry are sketched. Particular emphasis is placed on the role that both ideas about symmetry and symmetry breaking (this latter particularly in the Soviet Union), and ideas born of hadronic string theories (particularly in the West) played in pointing the way to supersymmetry. The related role that supersymmetry later played in reviving interest in string theories is also discussed. Although this brief account is necessarily somewhat narrowly focused, an attempt is made to view these developments from a long-term perspective. A study of the history of supersymmetry provides an ideal opportunity to obtain such a view, because supersymmetry has been at the center of intense interest during a time when both the theoretical and experimental study of high energy physics have changed in significant ways.

[a]This paper was prepared in 1987 for a conference proceedings which have never been published. The paper circulated as SUNY (Stony Brook) preprint ITP-SB-8878.

In 1987 a conference on the history of gauge field theories was held in Logan, Utah. The paper that follows is the paper I wrote as a contribution to the conference proceedings. The reason I agreed to give the talk, and then spent a good deal of time writing the paper, was that the seeds of interest in the history of science had been planted while I was a student at the Institute for Theoretical Physics at Stony Brook during the years after supergravity was invented. We were all excited about supersymmetry and supergravity in those days, and students had a chance to attend talks by just about every major figure in the West. In addition, because so many of these researchers made extended visits, we had a chance to talk with them, sometimes at length. Thus, we had a wonderful opportunity to witness the flow of work in supersymmetry, the rise of some ideas and the fall of others. This is the story I have tried to tell in the paper. I focused on the years up to 1985 simply because I did not have more time to devote to the project.

I benefited tremendously from contact with some historians of science and technology while I wrote the paper and afterward. Sam Schweber was a mentor to me, Spencer Weart was a source of support and encouragement. I learned a good deal from faculty and students at MIT, where I was conducting my own research in astrophysics, and where I also decided to expand my horizons by learning more about how to study the history of science. Perhaps the most precious resources came from the supersymmetry researchers themselves, many of whom read various versions of the paper for me, and made comments. The fact that this rather tough and demanding group of people reacted in a way that was almost uniformly positive reinforced my impression that it had been a good idea to rely solely on the published literature for my historical paper. The interesting conversations we had made me feel, however, that it would also be useful to future historians of science to capture an oral history of these conceptual developments. My goal was to write a book that would combine a systematic oral history with the bare facts of the physics as written in the literature and summarized here. Many of the key contributors to the conceptual developments have already contributed to this larger project by being interviewed. The book is "in progress," although in an arrested state of development, because the demands of my own research have not left enough time to pursue much work on the history of supersymmetry, supergravity, and superstrings.

Since the proceedings of the Logan conference were apparently never published, I was pleased to hear from Gordy and Misha that they wanted to include my paper in this volume. I had not thought that anyone would remember this old work that had circulated as an ITP preprint some 12 years ago. As I read the other contributions I once again feel of the fire and excitement in this field and become aware of the need for an even more systematic and comprehensive history.

Rosanne Di Stefano
September 25, 2000

Contents

Introduction

Supersymmetry is a symmetry which relates fermions to bosons. In the years since its discovery it has proved to be a powerful symmetry which has provided:

- internal symmetries which allow particles of different spin to sit in the same irreducible multiplet;

- quantum theories with "miraculous" ultraviolet cancelations;

- theories of supergravity with convergence properties that are better than those of standard general relativity;

- candidates for theories in which all of the known forces are unified;

- consistently coupled spin-$\frac{3}{2}$ fields;

- insights into the positive energy theorems of general relativity;

- improved string theories;

- a possible connection between gauge and Higgs particles;

- a possible solution to the hierarchy problem.

This partial list of its successes makes it easy to understand why supersymmetry has played such an important role in theoretical physics during the last fifteen or so years. Figure 1,[b] in which the number of published papers classified under the heading of supersymmetry is plotted as a function of time, serves to illustrate the

---

[b]The data plotted in figures one through three are the number of papers listed under these categories in the subject index of *Physics Abstracts*. Each number represents the sum of the January – June and July – December figures. (These numbers have been estimated.) In July – December 1986 there were no entries under the heading of *string theory*, but the listings under *duality and dual models* were dominated by papers on superstrings and related topics. In January – June 1987 there were approximately 300 papers listed under the heading of *string theory*, and *duality and dual models* was still dominated by superstring related topics.

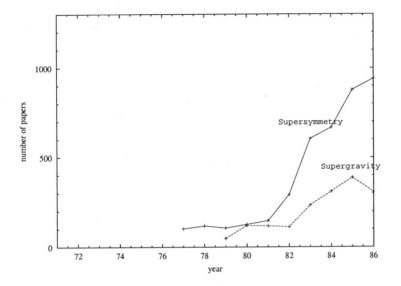

Figure 1: Supersymmetry and Supergravity

great deal of attention that has been focused on this area. An examination of the papers themselves reveals that the majority of them have been written by theorists. The enthusiasm of experimentalists has been more restrained. This is undoubtedly related to the fact that, as far as I know, the only data available to date on the existence of supersymmetry is data of the type plotted in Fig. 1.

In other words, one of the features that has characterized the development of supersymmetry so far has been the absence of any experimental evidence of its existence. The lack of experimental verification is not due to predictions of supersymmetry which have failed, but rather due to the fact that the crucial confrontations between supersymmetry and experiment have yet to take place. Nevertheless, this situation does put supersymmetry, at this point of time, on a different footing from many of the other symmetries which have been discussed at this conference.

To some extent, the idea of the possible existence of these other symmetries was inspired by physical observations. For example, the similarity between the proton and the neutron when electromagnetic effects can be ignored inspired the development of non-Abelian gauge

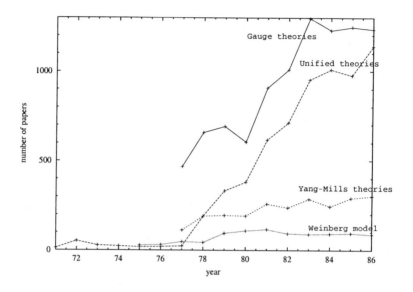

Figure 2: Field Theories

field theories. Or, the observed equivalence of inertial and gravitational mass was one of the elements that led Einstein to construct a theory invariant under general coordinate transformations. Theories with such invariances led to new predictions that were then verified by further experiments. In contrast, the history of supersymmetry thus far is the history of an idea.

One might imagine that the history of an idea would be rather tame. Nevertheless the story of supersymmetry contains all of the elements of high drama: an existence which defies all expectations, a double discovery, the birth of a new idea from an old and dying one which is then rejuvenated through the success of its offspring. The telling of this story will also provide an opportunity to explore what factors influence the generally perceived importance of an area of research, and to compare the research environments within which physicists have worked at different times during this century.

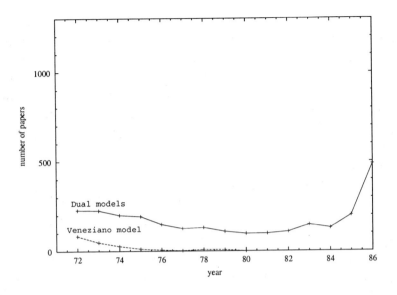

Figure 3: String Theories

Outline

The sheer volume and range of work that has been done on super-symmetry makes it impossible to include a comprehensive list of significant developments in a presentation of this length. Even if such a list, with a few descriptive remarks and the relevant attributions and dates, could be tailored to fit in this limited space, what purpose would it serve? Redundant to the specialist, not particularly enlightening to those uninitiated in its various subspecialities, it would not even begin to do justice to the dynamic nature of its subject. (Although the dynamic nature of the supersymmetries themselves could not fail to be noticed by anyone presented with such a list.)

On the other hand, if one were to attempt only an overview, there would be the danger of obscuring the details which make the field so rich. This presentation is an attempt to strike a balance. An overview will place supersymmetry within the larger tapestry of ideas that have been important in the evolution of theoretical physics over the past 65 years. On this tapestry, certain threads that are central to the overall pattern will be brought into sharper focus. Since some of the most interesting features of the pattern began to be studied in the earliest work on supersymmetry, some of this work will be discussed in detail. We will then follow a few of these individual threads, although in somewhat less detail, through subsequent developments. Because only a few threads can be studied in a review of this size, the reader who is not already familiar with the literature should be warned that many topics are not discussed here at all. For example superparticles and supersymmetric nonlinear σ-models have not been included, even though both of these areas have provided fertile ground for research. However, references to these, and to most of the other areas that are not touched upon here, can be found in the sources mentioned at the end of this section.

Section 1, on pre-superhistory, and Section 9, which briefly addresses the question of the future of supersymmetry, form a sort of historical frame which places supersymmetry within the broad context of high energy physics.

Section 2, on the dawn of supersymmetry, discusses the early work on supersymmetry which was done in the context of string theories of the strong interactions. The focus will be on the work of Ramond, and of Neveu and Schwarz, on dual models with graded algebras, and also on the discovery by Gervais and Sakita of the world-sheet supersymmetry of the Ramond-Neveu-Schwarz theory. This section, together with Section 8, which is a short section on the early work on superstrings, form a somewhat narrower frame. This more narrow frame will emphasize how the idea of supersymmetry was formulated in the context of dual string models of the hadrons, and how later, the study of supersymmetry led to consistent string theories as candidates for "theories of everything."

In the West, supersymmetry first started to be studied in 1971, in work on hadronic string theories. At that time it was hoped that such models might provide a good and useful description of the hadrons. The hadronic string theories were soon eclipsed by quantum chromodynamics. However the idea of supersymmetry lived on, largely because it was applied to local field theories in four dimensions by Wess and Zumino. The work of Wess and Zumino inspired a great deal of further work on supersymmetric field theories. Actually, supersymmetric field theories had already been constructed, and some of their properties had been studied, in the Soviet Union. Golfand and Likhtman, and Volkov, Akulov and Soroka had studied supersymmetric local field theories, but their work does not seem to have attracted much attention in the West. Thus, it was not until the independent work of Wess and Zumino, that supersymmetry came to be widely studied.

The question of what makes the time right for the study of any particular type of theory is an interesting one. Because of the way in which its development has been linked to other developments in string and field theory, and because it was discovered twice—once creating only a few ripples, and once creating a veritable tidal wave—supersymmetry provides a unique opportunity to explore possible answers to this question. Section 3, on the changing fortunes of string theories and local field theories, is an attempt to understand what made the time so right for the study of supersymmetric field theories in 1974.

Section 4 is on the dual discovery of supersymmetric field theories. Here the early papers by Golfand and Likhtman, by Volkov, Akulov and Soroka, and by Wess and Zumino are studied. The major themes that began to be explored after the work of Wess and Zumino are discussed. Of these, the unfolding understanding of the unique quantum properties of supersymmetric theories is studied in more detail in Section 5. Several sources are now available which provide detailed chronologies of the results which have been achieved. (See for example Ref. 132 and the contributions of Stelle and of West to Ref. 15.) The goal of Section 5 is to highlight the evolution of some of the important ideas and techniques.

One of the most important developments in the history of supersymmetry was the development of supergravity. The possibility that gravitation could be naturally formulated in the context of supersymmetry had been noticed, both by Volkov and Akulov, and by Wess and Zumino, in their earliest work on supersymmetry. Since the requirement of supersymmetry would naturally give the graviton at least one companion (either a spin-$\frac{3}{2}$ or a spin-$\frac{5}{2}$ field) and possibly other matter couplings as well, it was clear that supergravity might provide a natural path toward the unification of all of the know forces. As the cancelation of divergences which occurs in supersymmetric theories began to be studied, it appeared that supersymmetry might be able to overcome a barrier that had prevented the construction of a unified quantum theory—the barrier of the non-renormalizability of gravitation. It was hoped that supergravity theories (or at least one such theory) might be finite. This double possibility of unification and finiteness generated a great deal of excitement, and a great deal of creative work. The development of supergravity is outlined in Section 6.

One of the ideas that has played an important role in the progress of supergravity has been the Kaluza-Klein idea of achieving unification in higher dimensions. Of course, the notion of studying physics in higher dimensions had also of necessity been considered in the dual models, and we will discuss this connection as well. Section 7 is devoted to the study of supersymmetry in higher dimensions.

A Note on 'Conceptual Development'

It is difficult to define the notion of the 'conceptual development' of an area of research in a precise way. One is interested in knowing when key ideas were first developed and introduced, and also in understanding when they gained acceptance and became an established part of the foundation of the subject.

A survey of this kind cannot hope to record exactly when a particular idea might have occurred to, or might have been suggested by, a particular individual. Nor can it identify a point in time at which it became generally accepted that a particular approach might be important to the further evolvement of the subject. What it can do instead is to follow certain trends and broad ideas in the development of the subject, from the rough time that papers began to be published on these ideas. This survey is therefore only a step towards a more complete history of the subject. Because there has not been time to interview a broad enough cross section of the people who have participated in these developments, the published literature alone (in particular the dates on which manuscripts have been received by a journal) has been used to date the introduction of new ideas. Since much of the true infusion of these ideas occurred through lectures, discussions, and the distribution of preprints, the published record provides at best a first approximation to the timing of this process, but at least it is a well-defined approximation. To give a sense of the general perception of the progress of the field, these notes have relied on statements that were published at the time (for example in review articles) and also on statistics drawn from *Physics Abstracts* and *Current Contents.*

Because it was necessary to limit the space devoted to the list of references, this list surveys only a small portion of the relevant literature. Fortunately, many reviews, collections of reviews, and collections of pivotal papers are available. These would be useful to the reader who wants to go beyond the presentation sketched here. A collection of *Physics Reports* edited by Jacob[81] surveys the work on hadronic strings, while a collection of papers edited by Schwarz[115] deals with the first fifteen years of work on superstrings. Another useful source of information on string theories is the text by Green, Schwarz and Witten,[69] which also includes a comprehensive bibliography. Ferrara has recently edited a collection of papers on supersymmetry and supergravity,[47] while a collection of *Physics Reports* on supersymmetry has been edited by Jacob.[82] Texts include contributions by Wess and Bagger,[131] Gates, Grisaru, Roček, and Siegel,[57] and West.[132] A commentary on the history of Kaluza-Klein theories is included in the collection of papers on that subject which has been put together by Appelquist, Chodos and Freund.[6] Finally, the book by Abraham Pais, *Inward Bound*,[101] is an excellent source of information on the pre-supersymmetry background that is touched upon in Sec. 1.

1 Pre-Superhistory

1.1 Spin and Statistics

The idea of a fermi-bose symmetry presupposes a clear knowledge of quantum spin and of the relation between spin and statistics. These fundamental notions were developed during the 1920's. The research environment of the physicists who invented these ideas was quite different from that of the pioneers of supersymmetry. The differences are clear to physicists who work in today's environment, and who have studied something of the history of quantum theory. The contrast is highlighted by some of the facts and remarks which have been gathered together by Pais in his book. The quotations mentioned in this sections, as well as more details about developments that occurred before 1970 or so, can be found there.

One of the most important differences is the relation between experiment and theory. In the early days of quantum theory, experimental results forced the development of theoretical constructs. Commenting on the history of the Zeeman effect during the twenty fifth anniversary of its discovery, Lorentz said in 1921 that "theory could not keep pace with experiment." Indeed it was his study of the anomalous Zeeman effect which led Pauli to his 1925 formulation of the exclusion principle. When discussing his and Uhlenbeck's discovery of spin, Goudsmit remarked that the pair had been "saturated with a thorough knowledge of the structure of atomic spectra."

These comments serve to emphasize the mutual feedback and interplay between theory and experiment. There were unexplained experimental results which required the development of new ideas. And the consequences of those ideas, and further conjectures based on them, were then amenable to quick experimental tests.

This meant that the theories of those days were vulnerable to the discomfort of being proven wrong. Pais has compiled a list of conjectures which include: Einstein's suggestion in January of 1925 to apply Bose-Einstein statistics to electrons and to helium; Fermi's use, in March of 1926, of helium to illustrate the Fermi-Dirac equation of state; Dirac's conjecture of August 1926 that "the solution with antisymmetrical eigenfunctions... is probably the correct one for electrons in an atom and one would expect molecules to resemble electrons more closely than light quanta." Before the end of the decade, analysis of data on the specific heats of molecular hydrogen, and Wigner's proof that a gas of systems, each containing N fermions, obeys Fermi-Dirac statistics if N is odd, and Bose-Einstein statistics if N is even, showed that at least some of these conjectures were wrong.

On the other hand, access to experimental results gave theorists the ability to correct course quickly. This meant that they could afford to be wrong, without necessarily devoting a significant portion of their careers to work which was not immediately relevant to forming an understanding of the physical world. Furthermore, even if the results contradicted their guesses or predictions, they provided valuable guidance for the direction of further work.

The relation between physics and mathematics was also simpler in the 1920's. Although the analysis of quantum systems introduced many physicists to mathematics that was new to them, at least some of this mathematics was relatively accessible and was related to concepts that had already been used in the study of other physical systems. For example when, in 1926, Wigner introduced group theory to quantum theory during the course of his study of a system of N identical particles, he was using mathematics that he had worked with before. He had already used classical space groups while working on a chemical engineering problem on the lattice structure of rhombic sulfur.

1.2 The Role of Symmetry in High Energy Physics

From the early days of quantum mechanics and the work of Wigner, group theory played an important role in the study of atoms. Internal symmetry groups started to become important in the study of nuclei from the early 1930's when Heisenberg introduced the concept of isospin. (A brief history and references can be found in Ref. 135.) In 1937 Wigner introduced $SU(4)$ symmetry as a way to extend the $SU(2)$ symmetry group associated with isospin. He wrote: "To the extent that the strong forces inside the nucleus depend on neither orientation of spin nor isospin, they obey a stronger symmetry, $SU(4)$."

As the domain of experimental high energy physics expanded to include ever high energies, more particles were discovered. With the discovery of strange particles, the search for a hadronic symmetry which could embrace more than isospin, became more urgent. By 1964, when the discovery of the Ω^- confirmed a prediction of Gell-Mann which had been based on $SU(3)$, $SU(3)$ was widely recognized to be such a symmetry. At this point $SU(6)$ was introduced into high energy physics, to play the same role for $SU(3)$ that Wigner's $SU(4)$ had played for isospin. This work attracted a great deal of interest, and was successful in describing the hadronic spectrum. It also helped lead to the development of the no-go theorems. These will be discussed in the next subsection.

The symmetries discussed above were understood as global symmetries. In 1954 Yang and Mills suggested that the notion of symmetry (specifically the isospin symmetry) be expanded to include internal symmetry transformations which can be made independently at each space-time point.[135] The non-Abelian gauge theories that were born of their idea have been central to the role that symmetry has played in high energy physics. These Yang-Mills theories have been discussed in more detail at this conference; we will come back to them here in the section on field theory and, of course, when discussing supergravity.

1.3   No-go Theorems

The work on SU(6) inspired an interesting line of questioning: to what extent was it possible to combine internal and space-time symmetries in a non-trivial way? For example, could different particles in the same multiplet have different mass, spin, and internal quantum numbers? The consideration that seems to have been most cited during the mid 1960's had to do with the mass, and was motivated by the observed mass differences between particles in the same multiplet of a symmetry group I, like the hadronic SU(3). Were these mass differences necessarily due to symmetry breaking? Or was it possible that there was a larger group G, with respect to which the world was exactly symmetric, and which could encompass both I and the Poincaré group, P, in a non-trivial way? If so, then the generators associated with I and P would not necessarily commute, $[I, P] \neq 0$, and masses of particles within the multiplets associated with I could therefore be different.

In 1964 W.D. McGlinn studied this question for a special case in which I is a semisimple Lie group.[89] McGlinn showed that some general assumptions (specifically that the structure constants are antisymmetric, that the Jacobi identity holds, and that $[I, L] = 0$, where L represents the generators of the Lorentz group) are sufficient to ensure that $[I, P] = 0$. This was the first of a series of so-called no-go theorems. (See Refs. 16, 98, 99, 100 and references therein.) The last and most powerful of these was a theorem that was presented by Coleman and Mandula in 1967.[17] In a paper titled '*All possible symmetries of the S-matrix*', they were able to extend the no-go result to the case in which the internal symmetry group is an infinite parameter group. In order to delineate those conditions under which a non-trivial combination of space-time and internal symmetries is impossible, they made a set of well-defined assumptions about the one-particle spectrum. These included the assumptions that there are a finite number of particle types with mass less than any given finite mass, that the elastic scattering amplitudes are analytic functions, and that there is scattering.

There is no evidence that these no-go theorems prevented the exploration of useful avenues of research. In fact they played a positive role in that they gave a set of well defined situations in which the known symmetry groups could not be extended in the desired way.

Eventually these theorems were circumvented by considering systems with symmetry generators which participate in a (graded) algebra which includes anticommutators as well as commutators. An interesting side note is that in 1971 Lopuszański did consider the possibility of including fermionic generators among the symmetry generators of a theory.[88] He was working in the context of axiomatic field theory and his work seemed to indicate that such generators could not be included in a consistent way. In 1975, when supersymmetric theories had begun to be studied, it was Lopuszański who, with Haag and Sohnius, worked to determine the most general algebraic structures which include fermionic generators. The paper in which they presented their results was titled '*All possible generators of supersymmetries of the S-matrix*'.[76]

2   The Dawn of Supersymmetry

In the late 1960's, data about the hadronic spectrum suggested a theoretical hypothesis about hadronic scattering amplitudes which was referred to as duality. In 1968, Veneziano was able to construct a scattering amplitude which satisfied the duality hypothesis. This inspired a great deal of further theoretical work on dual models.

In 1970 it was discovered that the dual models could be viewed as models of hadronic strings. These hadronic string theories were found to have a number of interesting features. Some of these features are so attractive and compelling that they are even now drawing a new generation of physicists into the study of string theories as possible 'theories of everything'. There were also some problems. Some of these have since been worked out. Others are not problems when the arena for the string theories is on the order of the Planck length, rather than on the order of the size of a nucleon. One of the problems with the early theories was the fact that they could describe only bosons. Other problems included the presence of tachyons in the spectrum and a consistent formulation in only 26 space-time dimensions.

It was in the study of hadronic string theories that algebras which include both anticommutators and commutators were first used as symmetry algebras for high energy physical systems. (Graded algebras had already been studied by mathematicians. See Ref. 11 and references therein.) Such algebras were used by Ramond in an attempt to extend the dual theory formalism in such a way as to include fermions. They played an important role in the further development of string theories. Theories associated with such algebras proved capable of describing both fermionic and bosonic states. Such theories were found to be consistent in 10 space-time dimensions which, if not a particularly popular number for working physicists during that period, at least seemed preferable to the 26 dimensions necessary for such string theories to have a space-time supersymmetry and to be free of tachyons. (Here the word space-time refers to the full space-time arena in which the physics is taking place, as opposed to the two dimensional world-sheet that is swept out by the string.)

The first paper to derive a graded algebra for dual models was written by Pierre Ramond. It was almost immediately followed by important work by Neveu and Schwarz.

2.1   Graded Algebras

In 1971 Pierre Ramond extended the reach of dual models by inventing a way to incorporate fermionic states.[103] (Actually the work must have been done in 1970, for the paper was received by Physical Review on the 4th of January, 1971 and was published in the May 15th issue.)

His approach is remarkable, both for its ingenuity and for its simplicity. Ramond described the known bosonic states in terms of an internal time coordinate τ. He noted that this description had already been used by others. However he went on to define a physical limit in which the observables of the system can be written as averages taken over the period of the internal motion. For the free bosonic states this internal motion was assumed to be generated by the Nambu Hamiltonian,

$$H_B = \frac{1}{2} \sum_{n=0}^{\infty} [p^{(n)} \cdot p^{(n)} + \omega_n^2 q^{(n)} \cdot q^{(n)}], \tag{2.1}$$

where $\omega_{n+1} - \omega_n = \omega$, for $n = 0, 1, 2 \ldots$, and where the mode variables $q_\alpha^{(n)}$ and $p_\alpha^{(n)}$ are four vectors which satisfy bosonic bracket relations.

Ramond considered that the physical limit is achieved by taking the limit $\omega_0 \to \infty$. In this limit the observables were expressed as averages over the internal time coordinate. For example he defined the momentum of a bosonic state as

$$p_\mu = \langle P_\mu(\tau) \rangle. \tag{2.2}$$

In (2.2), $\langle P_\mu(\tau) \rangle$ represents the average

$$\langle P_\mu(\tau) \rangle = \frac{\omega}{2\pi} \int_{-\frac{\pi}{\omega}}^{\frac{\pi}{\omega}} P_\mu(\tau) d\tau, \tag{2.3.a}$$

and P_μ is defined to be

$$P_\mu = \sum_{n=0}^{\infty} p_\mu^{(n)}. \tag{2.3.b}$$

The Klein-Gordon equation was generalized: $p^2 - m^2 \longrightarrow \langle P \cdot P \rangle - m^2$. Solutions of this generalized equation were identified as the states of the free bosonic system.

To identify the fermionic states, he used the same procedure, but developed an equation which was analogous to the Dirac equation rather than the Klein-Gordon equation. Since the Dirac equation involves the Dirac matrices, Ramond first found generalization of these. He assumed that they should also be viewed as functions of the internal time coordinate, τ, with

$$\langle \Gamma_\mu(\tau) \rangle = \gamma_\mu. \tag{2.4}$$

By noting that the algebraic relations satisfied by the Γ's must reduce to the appropriate relations between the γ's when the averaging is performed, and by making the simplest assumptions which accomplish this, he was able to derive an explicit form for the generalized Dirac matrices:

$$\Gamma_\mu(\tau) = \gamma_\mu + i\omega_0 \tau \delta_\mu + i\sqrt{2}\gamma_5 \sum_{n=1}^{\infty} [b_\mu^{(n)\dagger} e^{i\omega_n \tau} + b_\mu^{(n)} e^{-i\omega_n \tau}]. \tag{2.5}$$

The important thing about this expression from the point of view of developing a fermi-bose symmetry is the presence of the b_μ^n's. These obey anticommutation relations.

$$\{b_\mu^{(n)}, b_\nu^{(m)}\} = \{b_\mu^{(n)\dagger}, b_\nu^{(m)\dagger}\} = 0, \tag{2.6}$$
$$\{b_\mu^{(n)}, b_\nu^{(m)\dagger}\} = -g_{\mu\nu}\delta^{n,m}, \qquad n, m = 1, 2\ldots$$

It is through these fermi-like coefficients, that anticommutators enter into the algebra of the dual models. To see how this works it is appropriate to pause in this development to see how an algebra is associated with a string theory in the absence of fermions. The treatment given below largely follows the development of Ref. 69 where the interested reader will find a detailed account.

Constraints in a theory often play two roles. They may serve as generators of symmetry transformations, and they naturally restrict the space of possible physical states. In bosonic string theories the energy-momentum tensor is constrained to be zero. Define L_m to be appropriately chosen Fourier components of the energy momentum tensor. The L_m are naturally also zero, i.e. they are themselves constraints. It is possible to define them in such a way that they satisfy the following classical algebra:

$$[L_m, L_n]_{P.B.} = (m - n)L_{m+n}. \tag{2.7}$$

This has been referred to as a classical algebra because the bracket relations are Poisson Brackets and not commutators. This algebra is the Virasoro algebra and the L_m are the Virasoro generators. They are symmetry generators because the transformations they generate are reparametrizations of the world sheet of the string, and such reparametrizations do not affect any physical results.

For the quantum system, the physical states $|\phi\rangle$ are those for which

$$L_m|\phi\rangle = 0. \tag{2.8}$$

We note that the L_n can be expanded in terms of harmonic oscillator-like creation and annihilation operators. Because of ordering ambiguities, for $m = 0$ the condition on the physical states must be permitted to be more general: $(L_0 - a)|\phi\rangle = 0$. Related to this is the fact that the quantum (commutator) algebra is slightly different from the classical one. It contains an anomaly term,

$$[L_m, L_n] = (m - n)L_{m+n} + \frac{D}{12}(m^3 - m)\delta_{m+n}. \tag{2.9}$$

This is a central extension of the Virasoro algebra.

The development above applies to the bosonic string. What Ramond did is to invent a generalization of the Dirac equation, using his generalized matrices,

$$[\langle\Gamma_\mu(\tau)P_\mu\rangle - m]|\psi\rangle = 0, \tag{2.10}$$

where $|\psi\rangle$ represents a fermionic state. He then defined Fourier components of the operator $\langle\Gamma \cdot P\rangle$,

$$F_{\pm n} = \langle e^{\pm i\omega_n \tau}\Gamma_\mu(\tau)P_\mu(\tau)\rangle. \tag{2.11}$$

With this definition, Ramond found that the algebra

$$[L_n, F_m] = \frac{1}{2}\omega(2m - n)F_{n+m},$$
$$\{F_n, F_m\} = 2L_{n+m} \tag{2.12}$$

is satisfied, where the L_n are generalizations of the Virasoro generators for the bosonic theory. Thus a graded algebra has emerged.

It was also possible to show, by using (2.10) and (2.12), that, in analogy to (2.8),

$$F_n|\psi\rangle = 0. \qquad (2.13)$$

Important to Ramond's investigation is that the spectrum of physical states was fermionic.

2.2   Dual Pion Models

Just a few months after Ramond's work was submitted, Neveu and Schwarz presented a paper in which they introduced a new dual model for pions.[94] The motivation and approach of their paper was quite different from that of Ramond, but their work makes contact with his in a very interesting way.

The goal of Neveu and Schwarz was to create a bosonic string theory with a more realistic spectrum. Toward that end they augmented the usual bosonic creation and annihilation operators with fermionic ones. The theory that they introduced was an interacting theory while, as (2.10) suggests, Ramond's work had applied to a free theory. In the course of their work they constructed generators G_r, where r is a half integer index, which participate in an algebra almost identical to the one derived by Ramond. This suggested that perhaps the fermi states of Ramond and the new bosonic states constructed by Neveu and Schwarz could be viewed as different states in a single model.

How this came to be understood takes us to the modern understanding of the space-time supersymmetry of superstrings, and so we will continue this discussion in Sec. 8. However at this point it may be appropriate to emphasize that the symmetries that were first understood to be present in the work of Ramond and of Neveu and Schwarz, were symmetries of the two-dimensional world-sheet of the strings. The nature of these world sheet symmetries was illuminated by the work of Gervais and Sakita.[58]

2.3 Strings with World-Sheet Supersymmetry

In string theories one can define fields which are functions of the coordinates of the two dimensional world-sheet. Since the world-sheet is itself defined in a space-time of higher dimension (for example 26 for the old bosonic strings and 10 for the theories discovered by Ramond, Neveu and Schwarz), these fields also have transformation properties on the larger space. For example the field Φ considered below is a space-time vector. However, in this section we will be concerned not with space-time, but rather with the world-sheet. For it was world-sheet supergauge symmetry which provided the most direct link between the work on string theories with a graded algebra, and the work of Wess and Zumino on supersymmetric field theory.

In their 1971 paper, '*Field theory interpretation of supergauges in dual models*',[58] Gervais and Sakita proposed to interpret the operators G_r, which had been introduced by Neveu and Schwarz, and later used by Neveu, Schwarz and Thorn,[95] as a generator of supergauge transformations. These transformations linked the field Φ, which is quantized using commutators, to fields ψ_i which were considered to be quantized using anticommutators. In order to write the transformations, Gervais and Sakita introduced an anticommuting c-number, χ,

$$\delta\psi_1 = \epsilon g_1(\partial_x \Phi)\chi,$$
$$\delta\bar{\psi}_2 = \epsilon g_2(\partial_{\bar{x}} \Phi)\chi,$$
$$\delta\Phi = -\frac{i}{2}\epsilon(g_2\psi_1 + g_1\bar{\psi}_2 + \bar{g}_2\bar{\psi}_1 + \bar{g}_1\psi_2)\chi. \qquad (2.14)$$

Under these supergauge transformations, and other (non-linear) supergauge transformations also introduced in Ref. 58, the Lagrangian associated with the functional integral of the theory transforms as a total derivative, indicating that the G_r can indeed be viewed as symmetry generators. These supergauge transformations are strikingly similar to the supergauge transformations later introduced by Wess and Zumino. However, the symmetry transformations that they generate are not space-time symmetries that present a non-trivial extension of the Poincaré algebra, and so do not themselves directly tread on the toes of the no-go theorems. It was left to the generalizations of these new symmetries to four-dimensional space-time to do that.

2.4   The Interplay Between High Energy Theory, Experiment, and Mathematics, circa 1974

This is a good point to make contact with the broader theme of how physics and the way it is done have evolved during the period around (and especially during) the era of supersymmetry.

There are both many similarities and many differences between the way that physics was done in the 1920's and 30's and the way which it was done by 1974. Perhaps the most obvious difference is the greater degree of specialization that existed in 1974, even among theorists. This specialization is responsible for the fact that most of the concerns that have been addressed by supersymmetry are the concerns of high energy physicists. And perhaps the most obvious similarity, one which may have been taken for granted at that time because it had been so central to the advance of science until then, was the relatively tight connection between theory and experiment. This relationship is perhaps most vividly illustrated by the discovery of the J/ψ in the autumn of 1974, the year that Wess and Zumino's first paper on supersymmetry was published. Its discovery was announced on he 11th of November, and ten days later the discovery of the ψ' was announced. Pais writes, "The general pandemonium following these discoveries compares only, in my experience, to what happened during the parity days of late 1956.... theorists delved in their grab bag of exotica. I recall early discussions of three options: it was a weakly interacting Z-like boson; or a resonance carrying free color, or charmonium".[101]

Although the discovery of what turned out to be charmonium was particularly exciting, at that time there was nothing unusual about this sort of close contact between theory and experiment. The ISR (interesting storage rings) had come on line at CERN in 1971, and Fermilab had begun operations at 200 GeV in 1972 and quickly achieved higher energies. Neutral currents were discovered in neutrino reactions in 1973.

It is clear that things were beginning to change though, because very high energies were needed to test some of the key predictions of theories like the electroweak theory and the new grand unified theories. As a result, the size and difficulty of the necessary experiments were increasing. One gets a feeling for the longer time delays between prediction and possible experiment from the example of the experimental discovery of the intermediate vector bosons of the electroweak theory, which did not occur until 1983, approximately 16 years after the formulation of the now standard electroweak model.

As for the sort of mathematics that high energy theorists were using, it had begun to change in many ways. In particular the more sophisticated mathematics that is associated with Yang-Mills theories was being mastered and used by physicists, and a great deal of interaction between mathematicians and physicists was associated with this process. Nevertheless, many of the notions that were still most useful to high energy theorists were related to the basic group theoretic concepts that were introduced to many physicists during the era of the development of quantum mechanics.

3   The Changing Fortunes of String Theories and Local Field Theories

Developments in local field theory and in string theory have generally not occurred in phase. When work on the hadronic string was cresting, interest in field-theoretic studies (having been buoyed by work on symmetry breaking, by the possibility of constructing a unified theory of the weak and electromagnetic forces, and by new results on the renormalization of Yang-Mills theories), was still rising. When quantum chromodynamics was generally accepted as a fundamental theory of the strong interactions, interest in hadronic strings plunged. Much later, when it began to seem unlikely that field theories with local supersymmetry would be found to give finite results, new results in string theory produced a surge of interest which seems to still be growing.

This section will be devoted to studying the relationship between the fortunes of local field and string theories over the period from 1970 or so through the present time. This evolving relationship has played a major role in the development of supersymmetric theories.

And supersymmetry has, in turn, provided an important and much used link between field and string theories.

The story begins in the late 1960's with efforts to understand the strong interactions. At that time there were concerns that the very strength of the interaction might make sensible application of a field theoretic perturbation scheme impossible. And not much was understood about non-perturbative approaches. Furthermore, hadronic resonances of high spin had been discovered, and it was known that there were problems of consistency in field theories that describe interactions of particles of higher spin, like spin-$\frac{3}{2}$ for example. The duality concept grew out of an attempt to understand hadronic data, and it incorporated the higher spin resonances in a natural way, It was discovered in 1970 that the dynamics of the theories that satisfy the duality hypothesis is the dynamics of a relativistic string.

The explorations which had led to the formulation of hadronic string theory had occurred at energies available to experimentalists at the machines that operated during the mid-1960's. By the end of that decade, scattering experiments at higher energies, which might have been capable of actually exploring the structure of the strings, had found evidence for point-like behavior instead. This behavior was inconsistent with the predictions of the string models, but could be described by a local field theory, quantum chromodynamics.

In the meantime, during the decade that spanned the early 1960's and the early 1970's, several important discoveries had been made about local field theories. In 1974 Wess and Zumino brought the idea of supersymmetry to the study of field theories in four dimensions. Their work, which was submitted to *Nuclear Physics* on the 5th of October 1973, triggered an avalanche of further work on supersymmetric field theories. Fayet and Ferrara[39] listed papers on supersymmetry that had been written by March 1977, when they updated the list of references in their *Physics Reports* during proof. They included approximately 135 papers on supersymmetry, which had been written by about 52 separate authors. In 1977 itself, 104 papers were listed under supersymmetry in *Physics Abstracts*.

Much of the excitement that was generated by these theories had to do with their remarkable ultraviolet properties, which were due to cancelations of divergences contributed by the fermi sector against divergences from the bose sector. These cancelations were of particular interest to theorists who had been struggling with the problem of quantizing gravity using perturbation schemes. Einstein's theory of gravitation had long been known to be non-renormalizable. This is to say that, if infinities appear order-by-order in the quantum perturbation expansion, then it is impossible to eliminate them while retaining the predictive power that a physical theory should have. Non-renormalizability need not be fatal if either the quantum expansion gives finite results in each order (so that renormalization is not needed), or if it can be shown that the sum of the terms in the perturbation expansion turns out to be finite. To test whether or not this latter possibility is realized would require the ability to do non-perturbative calculations for quantum gravity. To develop a workable approach to such calculations is, to say the least, extremely difficult. Although work has been done to develop non-perturbative methods, it would seem that we still have a long way to go toward a non-perturbative evaluation of the status of Einstein's theory as a predictive quantum theory. Most high energy theorists have preferred to study the perturbation expansion to see if the first hope, term-by-term finiteness, is realized. This is also a difficult enterprise. Although, in 1974, 't Hooft and Veltman were able to show that there are no physical problems for Einstein's theory in one-loop order, the calculations are so difficult that the two-loop result was not obtained until 1985. As many had long suspected, the two-loop result for pure gravity was not finite. Of course the calculations that would seem to be more physically relevant are those in which gravity is coupled to matter. Hopes that matter couplings might help to eliminate order-by-order divergences were dashed when, starting in 1974, possible couplings were shown to yield to divergent results even in one-loop order. This made the idea of incorporating supersymmetry into theories which include gravitation seem very attractive.

Fortunately, it proved possible to incorporate supersymmetry into gravitational theories by gauging the symmetry. For, when supersymmetry is a local symmetry, gravitation is automatically and necessarily present. Although the task of constructing and working with these theories of supergravity was not an easy one, such theories were first constructed in 1976 and a great deal had been understood about them by 1984. In particular the possible cancelation of divergences had been studied, and it had indeed been found that supersymmetry improved the situation vis-a-vis the divergences. The reason for the improvement had also come to be understood. However, although there was no direct evidence to suggest that supergravity theories could not be finite, it was not clear why they *necessarily* should be. In surveying the literature, one senses that by the early 1980's many of the experts had begun to doubt that even the maximally symmetric (N=8) theory of supergravity, which had provided the greatest hope, would be finite.

Even if the hope that supersymmetry would provide a predictive field theory of quantum gravity has apparently not so far been realized, supersymmetry has nevertheless made it easier to discuss quantum predictions of gravitational theories. This comes about because the presence of the symmetry makes the necessary calculations more manageable, and is illustrated by the fact that Grisaru was able to discover that pure supergravity is finite up to two-loop order, approximately eight years before a two-loop result was obtained for pure gravity. This tremendous increase in calculational power makes the incorporation of supersymmetry into theories of gravity still seem attractive.

If field theory was not to provide a predictive theory of quantum gravity, perhaps there was another way. An alternative was suggested by Scherk and Schwarz, way back in 1974. Their suggestion was based on the fact that field theories of gravity are necessarily plagued by ultraviolet divergences. The source of these divergences can be understood as follows. Although the Newtonian constant, G_N, has the dimensions of $(mass)^{-2}$, the terms in a probability amplitude must be dimensionless. Thus, those terms which are of order $(G_N)^n$ must include $2n$ factors of the momentum, and will therefore diverge at high energy. Because strings have finite extent, Scherk and Schwarz postulated that a string theory which includes gravitation should be able to avoid the ultraviolet divergences of the gravitational field theories. Although they and others did pursue this idea, only a few papers on it were published during the 1970's. One of the things that had begun to be understood, however, was the way that string theories could incorporate space-time supersymmetry.[59,60] This line of investigation was then systematically pursued by Green and Schwarz during the early 1980's, and by 1984 had begun to attract a great deal of attention. These more recent developments will be discussed in Sec. 8. The following subsections, on duality and local field theory respectively, give something of the flavor of the research that was being done in each area during the early 1970's. The role that this state of affairs played in the development of supersymmetry, as well as the role eventually played by supersymmetry in both field and string theories, is also discussed.

3.1 Duality and Hadronic String Theories

The goal of the dual resonance models was an ambitious one. As stated by Fubini in his 1974 introduction to a collection of *Physics Reports* reprints,[81] "The final aim is to collect all particles with the same baryonic number in a single supermultiplet." Furthermore the conceptual and mathematical structure of these models was extremely rich. It was discovered that dual theories can be formulated as theories of relativistic strings, indicating that the dual theories were describing the physics of extended objects.

In his 1974 *Physics Reports* article, Veneziano paints a picture of how the dual theories were regarded. (This review is reprinted in Ref. 81.) "The impression one gains through discussion of this subject in the scientific community is that of considerable polarization in the attitude towards it: there is a group of enthusiastic theorists who consider the problem of strong interactions as practically solved (up to some 'technical' points) and a group of more sceptical physicists who can hardly follow what is going on, and, basically, do not see compelling physical motivation behind the dual approach to strong interactions."

As mentioned in the previous section, there were three main weaknesses of the dual resonance models: the early models described only bosonic states; a consistent quantum formulation seemed to require 26 space-time dimensions; and the presence of tachyons plagued the spectrum. Supersymmetry proved capable of affecting all of these. The introduction of graded algebras enabled the string theories to describe both bosonic and fermionic states. These improved string theories with world-sheet supersymmetry had a consistent quantum formulation in 10 dimensions, which must have at least seemed like a step in the right direction. And the projection of these theories which turned out to have space-time supersymmetry eliminated the tachyons.

In spite of the benefits that supersymmetry bestowed upon hadronic strings, for most physicists interested in understanding more about hadronic systems, the focus of attention shifted from string theories to QCD during the early 1970's. Since the reason for the shift did not have to do with any of the problems of string theories, to most people the supersymmetric 'cure' seemed irrelevant. However ideas drawn from duality did not fade from hadronic physics altogether. The bag model, which was also developed in 1974 (see Johnson in Ref. 116) was an attempt to create a predictive scheme which had some of the features of the dual models, but also features of the quark and parton models. The bag, like the string, was an extended, finite object. The potential energy associated with it was taken to be a constant times the volume of the bag, analogous to the situation with the string. The development of QCD also included string like structures, in the tubes of SU(3) color. In fact, a significant number of papers on duality continued to be published through the 1970's and into the early 1980's. The low point occurred in 1980 and 1981, in each of which slightly more than 90 papers were published. Most of these papers, at least up until the early 1980's, were focused on the role of strings in hadronic physics.

3.2   Strings as 'Theories of Everything,' and the Connection Between String and Field Theories

Even as interest in string theory seemed to dry up, the seeds of its renaissance in a more powerful form were planted. In a paper which was submitted to *Nuclear Physics* on the 14th of May 1974, Joel Scherk and John Schwarz suggested that string theories could describe both the hadronic and the non-hadronic world.[111] (They noted that this had also been suggested by Bardakçi and Halpern.[10]) One of the key elements of their proposal was that it is possible to derive field theories as the low energy limit of string theories. The discovery that this is so had been initiated by Scherk in a paper which was published in 1971.[110]

Scherk had noted that the so-called Feynman-like diagrams (FLDs) of the dual models had been derived as generalizations of the Feynman diagrams of the ϕ^3 field theory, and had studied the questions of whether and how the ϕ^3 theory could be extracted from the dual resonance model. As in the later superfield formalism, each line in a dual FLD represents more than one field. In the FLDs, each internal line represents the exchange of an infinite number of excited states lying along a Regge trajectory. Scherk expanded the amplitudes of the theory in powers of the slope parameter α'; this corresponds to an expansion in terms of the fundamental length scale of the theory. He then considered the $\alpha' \to 0$ limit, which should correspond to a low energy limit. Since in this limit the mass of the excited states becomes infinite, Scherk reasoned that the ϕ^3 theory should be recovered.

Because the slope is not the only parameter in the theory—for example, there is also the coupling constant g—there is more than one way to take the $\alpha' \to 0$ limit. Neveu and Scherk took the limit differently and found that the on-shell amplitudes in the tree approximation are those of the 10-dimensional Yang-Mills theory.[96] Yoneya considered the $\alpha' \to 0$ limit of the Virasoro-Shapiro model (with $g\sqrt{\alpha'}$ fixed), and found that it corresponds to Einstein's theory of gravity.[136,137]

The massless particles associated with the gauge fields of these theories had no obvious role in a theory of hadrons. However, in their pivotal 1974 paper, Scherk and Schwarz noted that if string theories were considered as theories of non-hadrons, as well as of the hadrons, then the "presence of massless particles becomes a virtue."[111] They proposed to take this view seriously and to consider the consequences. They wrote, "... it is tempting to speculate as to whether a dual scheme might also be appropriate for the non-hadronic world (including leptons, photons, gravitons, and other gauge fields). Obviously there is no empirical evidence of duality or Regge behavior in non-hadronic interactions. However, the idea of having string-like rather than point-like particles is so general that it might extend to non-hadrons as well, provided that the fundamental length (or, equivalently, slope) characterizing the non-hadronic world is much smaller that the hadronic value, so as not to conflict with the limits imposed by the successes of quantum electrodynamics." Indeed they found that α' had to be quite small (approximately 10^{-34} GeV^{-2}), in order for a non-hadronic string theory to yield the correct value of the electric charge and of Newton's constant.

Scherk and Schwarz pursued this idea and its implications. In 1975 they won an honorable mention in the contest sponsored by the Gravity Research Foundation, for their essay, '*Dual model approach to a renormalizable theory of gravitation.*' They observed that the good ultraviolet behavior of string theories might make a consistent and predictive theory of quantum gravity possible. The Einstein field theory limit might then be to the full string theory, as the V-A point interaction scheme is to the full electroweak gauge theory. (This essay is reprinted as paper #18 in Ref. 115.)

One aspect of the string theories which needed to be reconsidered if they were to be thought of as fundamental theories within which all of the forces, including gravity, could be unified, was the meaning of their definition in higher dimensions. The possibility of taking the extra dimension seriously was studied by Cremmer, Scherk, and Schwarz,[18,11,112] who, along with Englert, were among the first to introduce the notion of using higher dimensions to study supersymmetric theories.

Even though this early conjecture about string theories as theories of everything did not attract very many active converts, it did seem to attract a fair amount of attention and favorable comment. For example, Olive considered it important enough to include in his plenary report on dual models at the XVII International Conference on High Energy Physics, which was held in 1974. After remarking that "almost all the classical field theories can be obtained as special cases" of the dual models, he went on to say that "we have the tantalizing possibility that there may emerge a unified theory of all the interactions.[116]"

3.3   Local Field Theory

The years between 1964 and 1974 had been years of tremendous development for field theory. Work on spontaneous symmetry breaking showed that observations in which a symmetry is not obvious, might nevertheless be predicted by a Lagrangian with a "secret symmetry." The discovery of the Higgs mechanism for such theories showed that the particle associated with a Yang-Mills gauge field could have non-zero mass. This established that the associated force could have finite range, and thus that the weak and strong interactions might be described by a Yang-Mills field theory. It also made it possible to see how, even in theories which can be described by a Lagrangian which respects an internal symmetry group, it is possible for different particles in the same multiplet of the group to acquire different masses. A gauge theory (now the "standard model") was constructed for the unified electroweak interaction. The renormalization of Yang-Mills theories was understood and shown to work even in theories in which the symmetry is spontaneously broken. A Yang-Mills theory was developed for the strong interactions. This theory, QCD, was shown to exhibit asymptotic freedom. Grand unification of the strong and electroweak forces was on the horizon.

The list of some of the accomplishments that had been achieved through the study of local field theory within a ten year period, gives some idea of the excitement that must have been generated by the study of field theory. Abers and Lee commented that, after the 1971 work of 't Hooft on the renormalization of Yang-Mills theories with spontaneously broken symmetry, there was "an explosion of interest in the subject, and the study of spontaneously broken gauge theories has become a major industry among theorists."[1] The advent of supersymmetric field theories provided an opportunity to explore these themes in a different setting. And indeed this is exactly what was done by a fairly large and growing number of field theorists. Spontaneous symmetry breaking, both for supersymmetry and for other symmetries which may exist in a supersymmetric field theory, was studied and attempts were made to construct realistic theories (particularly for the electroweak interactions) which were globally supersymmetric. The question of whether the neutrino was a Goldstone fermion was studied in detail, and it was found to be highly improbable. It was found that supersymmetric theories had special ultraviolet properties, and that there typically were cancelations among divergences in quantum calculations. A new formalism, the superfield formalism, which was specifically suited to the study of supersymmetric theories was developed. By 1976 supergravity had been discovered and a whole new area began to be explored.

3.4   Mapping the Changing Fortunes

The graphs on Figs. 1, 2, and 3 are plots of the number of papers published in several relevant areas during the years 1971 through 1986. The data was taken directly from *Physics Abstracts*. Naturally such information can give only a general idea of the amount of activity in each area and of general trends. [Among its weaknesses are these: there is a time lag in that it registers when the publication was listed and not when the paper was actually written; it relies on a classification scheme which must wait to confirm sustained interest before creating new categories; it does not give an idea of the proportion of physicists working in these areas as the number of physicists changes.] Nevertheless it does give a sense of what areas are being actively explored, and a way of comparing activity in different areas during any given year.

The graphs clearly indicate that the rise of supersymmetry and supergravity coincides with a time of increasing interest in field theories, particularly theories with a local internal symmetry and theories in which different forces are unified. It also indicates that interest in dual models never completely died out.

Unfortunately these graphs cannot by themselves give a clear idea of the situation in the years 1973–1975, which form the central focus of the discussion above. Yang-Mills theories and gauge theories did not even have their own listings until later. Furthermore, the number of papers on unified theories, which seems to have been more or less constant until about 1977, is not necessarily a good indicator. This is because in the early 1970's this heading was dominated by papers that discussed unification in the sense of Einstein, i.e. a unification of gravity with the other forces, in particular with electromagnetism, a unification that most high energy physicists did not dare to approach until somewhat later.

Perhaps the clearest indication that something special had happened in the early to mid 1970's was the creation of new headings by *Physics Abstracts* from 1975 through 1977. 'Weinberg models' started to be listed separately in 1975. This means that there must have been a steady or increasing number of publications per year on electroweak unification schemes during at least the few years prior to this. Yang-Mills theories and gauge theories were each given a heading in 1977. In that year there were approximately 119 papers listed under Yang-Mills theories, and a whopping 465 papers listed under gauge theories. Interest in these theories did not spring suddenly from nothing, and so it is clear that the intensity of interest in these areas must have been building by 1974. In the meantime, papers listed under the heading of 'duality' were already in the midst of a gentle decline by 1974, with a more pronounced decline evident for the number of papers listed under the 'Veneziano model'.

4 The Dual Discovery of Supersymmetric Field Theories

Physics is an international venture. It was therefore with some hesitation that this section was divided into two parts, one to deal with the discovery of supersymmetric theories in the Soviet Union, and one to deal with its discovery in the 'West'.

Field theories invariant with respect to fermionic symmetry transformations which participate in a non-trivial extension of the Poincaré algebra, were first considered in the Soviet Union. A paper on this topic was written by Golfand and Likhtman in 1971,[62] about two and a half years before Wess and Zumino submitted their first paper on supersymmetry. Golfand and Likhtman explicitly emphasized the possible fermionic extension of the Poincaré algebra. They presented an algebra, symmetry generators for two multiplets of free fields, and a method for constructing interacting Hamiltonians which respect their fermionic symmetry, along with an example of an interacting theory. Although the algebra that they presented is somewhat different from what is now called the supersymmetry algebra,[c] and

---

[c] The superalgebra found by Golfand and Likhtman was in fact the standard

although their approach was not used in the later development of the field, it is clear that key elements of later work were there.

Volkov, Akulov, and Soroka also did work based on the existence of a fermionic symmetry. Their interest was primarily in symmetry breaking. A paper by Volkov and Akulov, with the intriguing title, '*Is the neutrino a Goldstone particle?*', was published in *Physics Letters* in 1973.[126]

None of this work seemed to attract much attention outside the Soviet Union. Nor does the subject seem to have caught fire within the Soviet Union, although discussions with other physicists were mentioned by Volkov and collaborators, and although a great deal of later work on supersymmetry was to be done by Soviet physicists. As a matter of fact, the *Physics Letters* article of Volkov and Akulov did not refer to the paper by Golfand and Likhtman.

If there is a puzzle, it is not only to understand why this work did not attract more attention, but also why it did not do so at a time which was so right for the development of supersymmetric field theories. One may also wonder what a history of supersymmetry written today would have been like had it not been for the work of Wess and Zumino, which did catch fire. Would there have been follow-up on the work of the Soviets which would have eventually excited worldwide interest in the subject? Would someone else have discovered or invented supersymmetric field theories? If so, what circumstances would have ensured that this work would have been widely noticed and built on?

These are questions without answers. Yet to consider them seriously may not be a frivolous exercise. For although there is no way to directly explore the past (much less a hypothetical past), it is important to understand what factors led to one body of work being more or less ignored, while another body of work based on the same concepts, drew the attention and further work of physicists around the world.

---

super-Poincaré algebra, as we know it now, written in a somewhat non-standard notation. Instead of using the Weyl Spinorial generators of the superalgebra which is customary at present, Golfand and Likhtman worked with the Dirac spinors. This is the reason why they had to introduce $(1 \pm \gamma_5)$ projectors. – Editors' note.

The only aspects of these issues that can even begin to be addressed, are the specific ones of why the early Soviet work was not widely studied, and why the work of Wess and Zumino was. And since it is easier to explore the reasons why a positive action was taken than the reasons why an action did not occur, most of our efforts here will be directed toward understanding the follow-up to the work of Wess and Zumino.

It seems obvious that the fact that much of the early Soviet work was first published in Russian, and that only one paper of that entire body of work was published in a journal that is widely read by members of the international physics community, must have been at least partially responsible for the lack of international response. That is why this section has been split into two subsections, of 'East' and 'West'. There may have been other factors, such as the slight difference in timing, and the ease with which papers can get lost among the huge number of papers published every year, but these will not be speculated on here. As for the success of Wess and Zumino, it seems clear from the considerations of the previous section that timing did play a crucial role. There are also other factors which may have been important. These involve both the style and content of their early work—specifically that they explored the central ideas one by one, that they started to study the special ultraviolet properties of supersymmetric theories, and that they almost immediately began collaborating with other physicists on the study of supersymmetry.

4.1 Supersymmetry Rises in the East

Golfand and Likhtman

In 1971 Golfand and Likhtman wrote down the first field theory which was built to be invariant with respect to a fermionic symmetry.[62] Their motivation is clearly stated: they noted that "only a fraction of the [possible Poincaré invariant interactions] is realized in nature. It is possible that these [physical] interactions, unlike others, have a higher degree of symmetry.... we propose... a special algebra $\mathcal{R}$, which is an extension of the algebra $\mathcal{P}$ of the Poincaré group generators."

They introduced fermionic generators W_α and $\bar{W}_\beta$, which they called generators of spinor translations. ($\bar{W} = W^\dagger \gamma_0$.) The commutator of W with the generators $M_{\mu\nu}$ of Lorentz transformations are such that W transforms as a spinor. The usual relations of the Poincaré algebra are assumed to hold for the $M_{\mu\nu}$ and for the P_μ, the generators of translations. The new relations are:

$$\{W, \bar{W}\} = \frac{(1+\gamma_5)}{2}\gamma^\mu P_\mu,$$

$$\{W, W\} = 0,$$

$$[P_\mu, W] = 0, \qquad (4.1)$$

where the square brackets represent commutators and the curly brackets represent anticommutators. Having presented this algebra, they then constructed two separate generators which satisfy it:

$$W_1^0 = \frac{1}{i}\frac{1+\gamma_5}{2}\int d^3x(\phi^*(x)\overset{\leftrightarrow}{\partial_0}\psi_1(x) + \omega(x)\overset{\leftrightarrow}{\partial_0}\psi_1^c(x)), \qquad (4.2)$$

and

$$W_2^0 = \frac{1}{i\sqrt{2}}\frac{1+\gamma_5}{2}\int d^3x(\chi^*(x)\overset{\leftrightarrow}{\partial_0}\psi_2(x) + A^\mu(x)\overset{\leftrightarrow}{\partial_0}\gamma_\mu\psi_2(x)). \tag{4.3}$$

The superscript zero on each generator indicates that, when this generator is substituted into the equations (4.1) of the algebra, the P_μ that appears on the right hand side of the first equation corresponds to the free theory of the fields which are included in each multiplet. Thus W_1^0 and W_2^0 are each supersymmetry generators, as is their sum.

There are two features of this work thus far which form an interesting contrast to the work which was later done in the West. The first is that the algebra above is not really the algebra that is usually referred to as the supersymmetry algebra.[d] Because of the presence of the $\frac{(1+\gamma_5)}{2}$ which multiplies P_μ in the algebra, it is a sort of chiral version of the supersymmetry algebra. In fact Golfand and Likhtman comment that the theories associated with their algebra cannot be expected to conserve parity. Indeed the title of their paper was '*Extension of the Algebra of Poincaré Group Generators and Violation of P Invariance*'.

---

[d]See footnote on p. 203

The other feature that immediately strikes someone who is already familiar with the later and now more standard approach initiated by Wess and Zumino, is the absence of auxiliary fields. As will be discussed in subsection two, the multiplets that Wess and Zumino constructed contained both dynamical fields and other (auxiliary) fields that have no dynamics associated with them. These auxiliary fields entered into the Lagrangian in such a way that a subset of the Euler-Lagrange equations of motion are just algebraic equations for the auxiliary fields, and can be used to eliminate the auxiliaries from the action without loss of generality. When the auxiliary fields are eliminated however, two things happen. The first is that the supersymmetry transformations become non-linear, theory-dependent functions of the dynamical fields. Secondly, the supersymmetry algebra closes only when the equations of motion are used, that is only 'on-shell'.

Golfand and Likhtman did in fact comment that the equation of motion must be used in order to reproduce the equations of the algebra by substituting the particular forms W_1^0 and W_2^0 into (4.1).

As for the theory dependence of the generators, it was just this that they proposed to use in order to construct interacting theories which are invariant with respect to the transformations generated by W. In fact, in this paper, Golfand and Likhtman presented a method to construct invariant theories. Their method was based on expanding the generators of the interacting theory in powers of a coupling constant g. In particular, W and the Hamiltonian H are so expanded, and in this form, they are substituted into the relation $[W, H] = 0$. Terms proportional to equal powers of g are equated. After this it seems that the equations of motion, $\dot{W} = [W, H]$, are used.

Although one example of an interacting theory which has been constructed this way is given in their paper, it does not seem as if this method was used by others. (A somewhat different Hamiltonian-oriented method for the construction of supersymmetric theories was used later.[34,35,36]) In fact, given what was later understood to be the importance of the idea presented by Golfand and Likhtman in this paper, i.e. the idea of expanding the notion of symmetry of a field theory to include transformations generated by fermionic generators, the lack of response to their work is mystifying.

Volkov, Akulov, and Soroka

On the fifth of December 1972, Volkov and Akulov submitted a paper, '*Possible Universal Neutrino Interaction*', for publication. In it they entertained the hypothesis that the neutrino is a Goldstone particle.[125] In a paper that was submitted for publication a little more than a month later, they discussed the attraction of considering Goldstone and gauge particles with half integral spin.[3] "In the theory of elementary particles, a particularly important role is played by particles whose existence is made necessary and whose interaction is virtually uniquely determined by the presence of symmetry. ... [these are] all those associated with gauge fields ... and also Goldstone particles that arise in systems with degenerate vacuum." In order to study gauge and Goldstone fields with half-integral spin, it was necessary to study groups that are non-trivial extensions of the Poincaré group. This requires an extension of the group parameters, which must now contain spinor parameters.

The space of the coordinates must likewise be extended. So, in addition to the space-time coordinates x_μ, spinor coordinates $\psi(x)$ are introduced. The transformations of these coordinates that are expected to be symmetry transformations can be viewed as an extension of a symmetry transformation of the equations of motion of a free neutrino:

$$\psi \longrightarrow \psi + \xi,$$
$$x_\mu \longrightarrow x_\mu + \frac{1}{4i}(\xi^\dagger \sigma_\mu \psi - \psi^\dagger \sigma_\mu \xi), \qquad (4.4)$$

where the σ_μ represent the relativistic Pauli matrices.

This is both strikingly similar to and strikingly different from the superfield approach which was later introduced by Salam and Strathdee, and which is described in more detail in the next section. It is similar, in that in both approaches the coordinates of the four dimensional space-time are augmented by a set of spinor coordinates. It is different, in that Volkov and Akulov give a physical interpretation to the spinor coordinate, which is assumed to be a function of the space-time coordinate x_μ. $\psi(x)$ plays the role of the Goldstone field. In the superfield as first introduced by Salam and Strathdee,[105] the spinor coordinates are multiplied by functions of x_μ, and it is these functions that are identified as the physical fields.

There are other differences between this work and most of the later work on supersymmetry, and this is related to the point made above. In the later work on supersymmetry, a supersymmetry transformation generally related fermi fields to bose fields. On the other hand, in this early work of Volkov and Akulov, the only physical field was the spinor field ψ. Furthermore the transformation of each of the fields was later generally expressed as a linear transformation. Because of the x dependence of Volkov and Akulov's ψ, its full transformation is inherently nonlinear. Although nonlinear transformations have been considered in later work, this was the exception and not the rule, especially for the early work on supersymmetry in the West.

In their work on Goldstone fermions, Volkov and Akulov introduced differential forms:

$$\omega_\mu = dx_\mu + \frac{a}{2i}(\psi^\dagger \sigma_\mu \psi - d\psi^\dagger \sigma_\mu \psi). \tag{4.5}$$

These were used to construct an invariant action,

$$S = \int |V| d^4 x,$$

where

$$V_{\mu\nu} = \delta_{\mu\nu} + a T_{\mu\nu},$$
$$T_{\mu\nu} = \frac{1}{2i}(\psi^\dagger \sigma_\mu \partial_\nu - \partial_\nu \psi^\dagger \sigma_\mu \psi). \tag{4.6}$$

The symmetry determined the possible interactions of the Goldstone particle.

There were formal considerations which seem to have led Volkov and Akulov to their idea of a Goldstone fermion. These can be understood from the text of their papers, and also from a paper by Volkov on phenomenological Lagrangians.[124] This latter, seems to have been published close to the time that Volkov's work with Akulov was done. Once the idea proved to be theoretically feasible, they began to consider its phenomenological implications. For example in Ref. 4 they discussed the possible role of Goldstone fermions in astrophysics. "At a definite stage of stellar evolution, when there is a high density of high energy photons, the universal interaction of the Goldstone neutrinos can ... play the fundamental role in the cooling of the star. It would be if interest to consider the influence of this interaction during the early stage of expansion in the hot universe model."

Volkov and Akulov ended their first paper on supersymmetry[125] by noting that: "The gravitational interaction can be introduced into the scheme in analogous fashion, by introducing gauge fields corresponding to the Poincaré group. ... If we introduce also gauge fields corresponding to the [fermionic] transformations, then, as a consequence of the Higgs effect, a massive gauge field with spin-$\frac{3}{2}$ arises, and the Goldstone particles with spin-$\frac{1}{2}$ vanish." This seems to be the first suggestion to consider a supersymmetric theory of gravity. It is also the first mention of the possibility of the phenomenon which is now referred to as the super-Higgs effect.

The generalization to a local symmetry, and the super-Higgs effect, was studied by Volkov and Soroka in Ref. 127. This was accomplished by generalizing the differential forms to include the associated gauge fields, and then constructing invariants. The theory which results from such a procedure includes gravitation, a spin-$\frac{3}{2}$ field, a spin-1 Yang-Mills field, and, in cases in which the super-Higgs effect doesn't hold, a spin-$\frac{1}{2}$ fermion.

Although the work of Soviet authors does not seem to have been widely known in the West until after the independent discovery of supersymmetry had occurred there, it did eventually receive attention, and does seem to have influenced some of the later work on supersymmetry. For example, the question of whether or not the hypothesis that the neutrino is a Goldstone fermion is consistent with what is known about the phenomenology of neutrinos, was later studied in the West.[49] (This will be briefly discussed in Sec. 9.) The work of Volkov and Soroka on the inclusion of gravity was cited by Zumino in his talk at the *XVII Conference on High Energy Physics*.[116] Zumino later went on to try to construct a physical theory of supergravity in the arena of what he called the "curved superspace" introduced by Volkov and Soroka. Arnowitt and Nath also did a great deal of work on this. This work will be sketched in Sec. 6.

The Western work on curved superspace seems to have then, in combination with the early Soviet results, sparked further efforts in this direction by Volkov and his collaborators.[5] Likhtman also followed up on the work that he had done with Golfand, in a way that indicated that he was familiar with the Western literature on supersymmetry. He considered questions of renormalizability and also used superfield techniques.[86,87] And so, as these examples demonstrate, a genuine scientific exchange did begin between 'East' and 'West', thus obviating the need for further separate discussions after the following subsection on the early work of Wess and Zumino.

4.2 Supersymmetry Rises in the West

Wess and Zumino

In 1973 Wess and Zumino made the step of transplanting supergauge transformations from the seemingly dying body of dual models to the quickly growing body of field theories in four dimensions. Clearly this was the right step, taken at the right time. Their first paper on the subject was received by *Nuclear Physics* on the 5th of October 1973. Within two years tremendous progress had been made, both by Wess and Zumino and by others, toward understanding the remarkable features of supersymmetric theories.

In less than a year, three papers were published by the Wess-Zumino team. Each one focused on a particular aspect of supersymmetric field theories.

The first paper, '*Supergauge Transformations in Four Dimensions*', was devoted to studying the supersymmetry algebra.[128] Two multiplets of fields were introduced and a set of supergauge transformations was given for each multiplet.

The scalar multiplet consisted of two scalar fields $A(x)$ and $F(x)$, two pseudoscalar fields $B(x)$ and $G(x)$, and a Majorana spinor field $\psi(x)$. As did the supergauge transformations of Gervais and Sakita, these transformations connected bose and fermi fields,

$$\delta A(x) = i\bar{\alpha}\psi(x),$$
$$\delta B(x) = i\bar{\alpha}\gamma_5\psi(x),$$
$$\delta\psi(x) = \partial_\mu(A(x) - \gamma_5 B(x))\gamma_\mu\alpha + (F(x) + \gamma_5 G(x))\alpha$$
$$\qquad + n(A - \gamma_5 B)\gamma^\mu\partial_\mu\alpha,$$
$$\delta F(x) = i\bar{\alpha}\gamma^\mu\partial_\mu\psi(x) + i(n - \frac{1}{2})\partial_\mu\bar{\alpha}\gamma^\mu\psi,$$
$$\delta G(x) = i\bar{\alpha}\gamma_5\gamma^\mu\partial_\mu\psi(x) + i(n - \frac{1}{2})\partial_\mu\bar{\alpha}\gamma_5\gamma^\mu\psi(x), \qquad (4.7)$$

where the α_i are totally anticommuting spinors, and where n is arbitrary.

The canonical dimensions of F and G can be read off from these transformations, and they are clearly not those appropriate for conventional dynamical scalar or pseudoscalar fields. F and G are auxiliary fields. When supersymmetric Lagrangians were constructed, these fields appeared in such a way that they could be eliminated. However, their presence ensured that the supergauge transformations were linear, theory-independent and, as remarked upon by Wess and Zumino in this paper, was essential for closing the algebra (at least when the equations of motion are not satisfied).

The closure of the algebra was studied by computing the commutator of two successive transformations. If the α_i are not constant, then the commutator of two supergauge transformations involves a conformal transformation, a mixing of the spin-0 fields, and a γ_5 transformation of the spinor field. Because of the problems of anomalies associated with these symmetries, Wess and Zumino limited their attention to constant spinor parameters in their later work; however in this paper they did work out the full algebra, and also verified the Jacobi identities. The algebra that was later referred to as the supersymmetry algebra is the one that is obtained by requiring the spinor transformation parameters to be constant. In this case the result is

$$[\delta_1, \delta_2]\mathcal{F} = 2i\xi^\mu\partial_\mu\mathcal{F} \qquad (4.8)$$

for all the fields $\mathcal{F}$, where $\xi^\mu = 2i\bar{\alpha}_i\gamma^\mu\alpha_2$. This relation indicates that the generators of translations play a non-trivial role in the supersymmetry algebra, and thus that the algebra is a non-trivial extension of the Poincaré algebra. Equation (4.8), together with the conditions that the Hamiltonian and translation generators be supersymmetric, and the condition that Q transforms as a spinor under Lorentz transformations, defines the algebra of simple supersymmetry. The later work of Haag, Lopuszański, and Sohnius studied the possible extensions of this algebra.[76e]

Equation (4.8) also holds for the vector multiplet, which Wess and Zumino introduced in this first paper as well. The vector multiplet consists of the fields v_μ, λ, D, C, M, and N, which are subject to the supergauge transformations,

$$
\begin{aligned}
\delta v_\mu &= i\bar{\alpha}\gamma_\mu\lambda + i\bar{\alpha}\partial_\mu\chi, \\
\delta\lambda &= [-\frac{1}{2}F_{\mu\nu}\gamma^\mu\gamma^\nu + D\gamma_5]\alpha, \\
\delta\chi &= [\gamma^\mu v_\mu - \partial_\mu C\gamma_5\gamma^\mu + (M + \gamma_5 N i)]\alpha, \\
\delta D &= i\bar{\alpha}\gamma_5\gamma^\mu\partial_\mu\lambda, \qquad \delta C = i\bar{\alpha}\gamma_5\chi, \\
\delta M &= i\bar{\alpha}\lambda, \qquad \delta N = i\bar{\alpha}\gamma_5\lambda,
\end{aligned}
\tag{4.9}
$$

where $F_{\mu\nu} = \partial_\mu v_\nu - \partial_\nu v_\nu$, and we have assumed that α is constant. Another important part of this paper was the discovery of a way to construct scalar Lagrangians by combining multiplets. This method uses the form of the transformations of the auxiliary fields, for example δF in (4.7) or δD in (4.9), which are simply total derivatives. By combining multiplets in such a way that the combinations themselves transform as the fields of a multiplet do, and by inventing combinations for which the expressions corresponding to auxiliary fields are interesting enough to themselves be Lagrangians, one can construct theories for which the action is invariant up to a surface term. Wess and Zumino used this method to find that the Lagrangian of the free fields of the scalar multiplet corresponds to the auxiliary field D of a composite vector multiplet,

$$
\mathcal{L}_s^0 = -\frac{1}{2}[\partial_\mu A\partial^\mu A + \partial_\mu B\partial^\mu B + \bar{\psi}\gamma^\mu\partial_\mu\psi] + \frac{1}{2}[F^2 + G^2].
\tag{4.10}
$$

---

[e]Extended superalgebras were first discussed by A. Salam and J. Strathdee, *Nucl. Phys.* B **80**, 499 (1974). – Editors' note.

Furthermore, the free Lagrangian of the vector multiplet is proportional to the auxiliary field F of a scalar multiplet,

$$\mathcal{L}_v^0 = -\frac{1}{4}F_{\mu\nu}F^{\mu\nu} - \frac{i}{2}\bar{\lambda}\gamma^\mu\partial_\mu\lambda + \frac{1}{2}D^2. \qquad (4.11)$$

In this first paper Wess and Zumino particularly stressed the significance of the [complete] algebraic structure that they had developed. "The fact that the conformal algebra in four dimensions can be extended to the algebraic structure described above... and that the entire algebraic structure is generalized by the supergauge transformations, does not appear to have been realized before. It is interesting in itself, irrespective of the field-theoretic applications that we have in mind." They also noted that one may expect, by generalizing their supergauge transformations, to obtain an algebra which includes general coordinate transformations.

In their second paper, Wess and Zumino presented an interacting theory of the fields of the scalar multiplet.[129] After the auxiliary fields are eliminated, the Lagrangian is

$$\mathcal{L} = \mathcal{L}_s^0 - \frac{1}{2}[m^2(A^2 + B^2) + im\bar{\psi}\psi] \qquad (4.12)$$

$$- gmA(A^2 + B^2) - \frac{g^2}{2}(A^2 + B^2)^2 - ig\bar{\psi}(A - \gamma_5 B)\psi.$$

In principle, there could have been 10 independent parameters in such a theory (assuming that only terms consistent with parity conservation and renormalization are included). The fact that the requirement of supersymmetry allows only two independent parameters illustrates one of the most striking features of supersymmetric theories.

Wess and Zumino did a one loop calculation and found that renormalization did not force the introduction of new parameters, and later work by Iliopoulos and Zumino showed that the relationship between the parameters holds for the fully renormalized theory.[80] This stability of the relations between parameters plays an important role in phenomenological applications of supersymmetry. For example, it helps to maintain the huge separation between hierarchies in those grand unified theories which incorporate supersymmetry. Once they are set, they are not subject to radiative corrections in super-GUTS theories.

Related to this is the fact that, in this second investigation, Wess and Zumino discovered the special renormalization properties of supersymmetric theories. They especially noted the cancelations of divergences among different diagrams, cancelations which made the theory easier to renormalize than the generic theory of this form that can be built from this multiplet of fields. These properties will be the focus of the next section. They ended the paper by looking toward the future. "The present investigation suggests two interesting lines of research. The first is the study of higher order corrections. The second is the construction of more complex and hopefully more realistic models, invariant under a combination of supergauge and internal symmetries." Within a year both lines had been followed, both by Wess and Zumino and by others. [See Ref. 39 for a more complete list of references on this further work that the one compiled at the end of this paper.]

In their third paper, Wess and Zumino studied the interplay between gauge symmetry and supersymmetry. They constructed a supersymmetric gauge theory, a "supergauge invariant extension of electromagnetism".[130]f Two symmetry transformations, performed successively, produced a sort of generalized gauge transformation.

First a Lagrangian symmetric with respect to both the gauge and supergauge transformations, and hence also under the generalized transformation, was constructed. Because this form was so unwieldy (an infinite power series in the coupling constant), a special gauge was chosen, so that in its final form only the gauge, and not the supergauge symmetry was manifest. In this form, the only fields from the vector multiplet that survived were v_μ, λ, and D. For this theory too, there was just one mass parameter and one coupling constant. Again this property was preserved in the one loop approximation, and again there were cancelling divergences.

Although these few paragraphs cannot completely summarize their work, it is clear that the early work of Wess and Zumino formed

---

fIn fact, supersymmetric QED was first constructed by Golfand and Likhtman, Ref. 62. We suggest to call this model the Golfand-Likhtman model. – Editors' note.

a step by step approach to understanding some of the important features of supersymmetric theories. In particular they developed the algebraic structure, successfully constructed interacting theories, and had begun to study the special ultraviolet properties of supersymmetric theories. They also constructed a gauge theory and explored the relationship between supersymmetry and gauge symmetry. It is also clear that by this point they had discussed their work with a number of active physicists, some of whom had already begun to work on supersymmetry themselves. In these papers they mentioned discussions with J.S. Bell, S. Coleman, J. Iliopoulos, and E. Remiddi, in addition to B.W. Lee. In their third paper they were able to refer to other work on supersymmetry by Fayet, Iliopoulos, Salam and Strathdee.

Supersymmetry Takes off

The structure that Wess and Zumino had begun to uncover was interesting and highly relevant to other work which had been done in the study of field theory during the previous decade. Supersymmetry gave field theorists an opportunity to explore gauge theories, symmetry breaking, and renormalization in a new and attractive setting. The bypass of the no-go theorems also raised questions of how far one could go toward including generators of other symmetry transformations in the supersymmetry algebra. The cancelations of divergences presented the possibility of a finite quantum theory of gravity. The construction of supergravity raised new questions whose answers were eagerly pursued.

In 1977, a *Physics Report* about supersymmetry was published by Fayet and Ferrara.[39] In studying the references in that report, including the ones that were added in proof in March of 1977, one can see that all of these questions had been or were being considered. In many cases one can trace interest in these questions back to the pre-supersymmetry explorations of the authors. Some of the contributions to that early body of work on supersymmetry (although not a very large fraction) were from people who had worked on hadronic strings.

Certainly the most important reason for the rapid involvement of so many physicists in supersymmetry is the intrinsic interest of the subject. However, the fact that so many people were prepared to embark on the study of supersymmetry on such short notice is also related to the particular time at which the work by Wess and Zumino appeared. Had hadronic string theories proved capable of explaining the high-energy data, had the problems involving tachyons and higher dimensions been solved, had the calculations of gravity coupled to some form of matter given finite results, and had the electroweak theory already achieved the level of experimental verification that it now enjoys, it seems unlikely that the call of supersymmetry would have been so quickly and enthusiastically answered. It even seems possible that part of the reason that the earlier work of the Soviets was overlooked may have had to do with the small difference in timing.

Wess and Zumino continued to work on supersymmetry and published further work, both alone and in collaboration with others. If one looks at their collaborators over the years from 1974 through 1976, and the collaborators of their collaborators, one finds that they account for a large percentage of the early papers on supersymmetry.

5 Quantum Behavior of Supersymmetric Theories

Long before the invention of supersymmetry, it was realized that divergences associated with fermions might be expected to at least partially cancel some divergences associated with bosons. Given this, one can imagine that it would have been possible to discover supersymmetry in a way that is different from what actually happened. One can imagine that some physicist studying field theories which included both fermi and bose fields, some physicist who was perhaps tired of encountering divergences, would try to see if things could be arranged so as to let the divergences associated with different fields cancel among each other. This mythical physicist might have studied how the divergence structure of the theory depended upon the number of fields considered and upon their masses and coupling constants, and might then have seen how to minimize the divergences by allowing these parameters to be interrelated. Had such a path been taken, it might not have been obvious that the theory had an underlying supersymmetry. However the theories so derived would almost undoubtedly have been considered to be important, both because of their divergence properties, and because of the decreased number of arbitrary parameters.

Of course this is not the way things actually happened. But it is an interesting fact that something like this sort of bose-fermi cancelation seems to have been anticipated by Bohr. Bohr alluded to the possibility of such a cancelation in his lecture at the 1947 Cavendish Conference. And, in his lecture at the 1947 Summer Symposium at the University of Michigan, Pais mentioned Bohr's conjecture. "Bohr has suggested that . . . one might try to cancel the infinite negative energies of the Dirac seas (for electrons, protons, neutrons, and neutrinos) by the positive infinities of the zero point energies (of photons, and other Bose-Einstein fields, like mesons etc.) Thus, certain infinities would only be apparent because we are only considering one kind of elementary particle. Thus a deeper understanding might yield a connection between the two problems: the problem of the multitude of particles and the paradoxes of the infinities." (This is a quote from the notes of C.S.W. Chang.)

When Wess and Zumino introduced their first interacting theory, the first aspect of it that they explored was its renormalizability. Their early studies of the quantum properties of supersymmetric theories were followed by other investigations that continued to find cancelations among divergences associated with different Feynman graphs. Indeed, one of the questions that was considered in the year or so after Wess and Zumino's first paper, was the Vacuum energy in supersymmetric theories. In his paper, '*Supersymmetry and the Vacuum*',[138] Zumino found that ". . . the sum of all vacuum diagrams vanishes identically as a consequence of compensations among contributions among different fields of the supermultiplet. This means that the zero-point energy-momentum density is zero to all orders in the supersymmetric interaction."[g]

With Salam and Strathdee's introduction of superfields,[105] it became possible to not only construct supersymmetric theories with

---

[g]In fact, this property was first noted in the supersymmetric context by E. Likhtman in 1971 (Lebedev Physics Institute Report # 41, in Russian). — Editors' note.

more ease, but also to perform quantum calculations in a way in which these cancelations could occur automatically. Supergraph techniques to perform these quantum calculations were developed and refined over a number of years. Their power, especially when formulated in an improved form, was demonstrated by a series of so-called non-renormalization theorems.

5.1   The Early Work Which Was Done Without Superfields

In their first paper on an interacting supersymmetric theory, Wess and Zumino commented: "It is remarkable that this Lagrangian appears to be renormalizable even when the masses and coupling constants are not independent and that the relations among them are preserved by renormalization. Furthermore the theory turns out to be less divergent when the [relations between the masses and the coupling constants] are satisfied." (At this point the authors included a note to thank B.W. Lee for "pointing out first the occurrence of the divergences and also the fact that relations among couplings are preserved in the one-loop approximation.")

In fact it was the relationships between the mass and coupling constants which were responsible for the good ultraviolet behavior. For these relations were such that divergent contributions from different Feynman graphs were able to cancel among themselves. The authors noted that: "The quadratic divergence of the self-energy cancels among the various diagrams..." They also observed that possible corrections to off-diagonal mass terms, and to the couplings as well, add up to zero. "All these cancelations among different diagrams are due to the supergauge invariance of the Lagrangian."

The result was that, at least in the one-loop approximation, only one wave function renormalization common to all of the fields, was needed.

In a paper which was submitted to *Nuclear Physics* about two and a half months later, Iliopoulos and Zumino extended this result to all orders, and proved that just a single wave function renormalization was needed to renormalize the theory.[80] Their paper places special emphasis on the mutual cancelations, and includes specific diagrams to illustrate the point. In studying the reason for the cancelations they observed that it was not merely the symmetry under supergauge transformations alone which was responsible for them, but also the specific form of the Lagrangian, which had been dictated by both the requirement of supersymmetry and the inclusion of only terms which could be expected to be present in a renormalizable theory.

Lang and Wess later tried to relax this latter condition to test whether the power of supersymmetry to soften divergences would be great enough to allow a theory which by naive power-counting would seem to be non-renormalizable, to be renormalized.[84] They considered a specific Lagrangian, constructed with the fields of the scalar multiplet. The Lagrangian that they considered includes terms which should have spoiled the renormalizability. Although supersymmetry did not prove to be capable of completely saving the situation, Lang and Wess did find that the requirement of supersymmetry softened the divergences of the theory.

The cancelations which made these results possible were sometimes referred to as 'miraculous'. It turned out to be possible to understand the reasons that supersymmetry forces such cancelations, and to more readily predict their occurrence by using the superfield formalism. In that formalism, a single entity, a superfield, describes a set of fields defined on the four dimensional space-time. These fields are referred to as the component fields of the superfield. They might correspond to the fields of the Wess and Zumino scalar multiplet for example, or to those of the vector multiplet. They generally include both the dynamical fields of the multiplet and the auxiliary fields. By considering diagrams which are associated with the propagation of its components, the cancelation of divergences can happen automatically. That is to say that, if the diagrams associated with the component fields have divergences which cancel among themselves, the superfield diagrams would simply give finite results.

These observations were used in early work which developed and used the superfield Feynman rules to do quantum calculations. Superfield methods have since evolved into even more powerful tools.

5.2 Superfields

Within months after the publication of Wess and Zumino's first paper on supersymmetry, Salam and Strathdee introduced the superfield, as a way to "construct at least some of the representations of this symmetry.[105]" The superfield turned out to be a powerful tool for the construction of multiplets, supersymmetric theories, and for quantum calculations.

The idea of the superfield used the totally anticommuting property of the elements of a Grassmann algebra in an ingenious way. The mathematics associated with Grassmann algebras was being used in the study of quantum systems by the late 1960's. The reference that was probably most widely quoted at that time was Berezin's book on second quantization which was published in the West in 1966.[11] Berezin worked in the Soviet Union, and his early work was first published in Russian. One can wonder whether the key role played by Berezin and his colleagues, both in the development of the mathematics and in its application to quantum theory, may have been a factor in the earlier sprouting of some supersymmetric ideas in the Soviet Union.[h]

The work of Berezin also received attention from physicists in the West, especially after the publication of his book in English.[11] Citing this book as a good introduction to the subject, Montonen introduced totally anticommuting quantities to string theory in 1973,[91] and this work was soon followed by a paper by Fairlie and Martin which also used Grassmann variables. Whether or not the original work of Salam and Strathdee has roots that go back to the work of Berezin and the work in string theories, it will be useful to briefly review some of this earlier work. This review will provide an opportunity to define the basic concepts and also to get a sense of the ideas that were being discussed in 1974. To avoid unnecessary confusion, the Grassmann variables will be denoted in the same way for all of the work discussed below, they will be represented by θ.

---

[h]The work of Berezin and Katz served as an important impetus for Volkov and Akulov, see Akulov's note in this volume. – Editors' note.

Grassmann Variables

In his book, Berezin considers a finite dimensional Grassmann algebra $\mathcal{G}_n$, with n generators $\theta_1, \theta_2, \ldots \theta_n$. The generators are totally anticommuting,

$$\{\theta_i, \theta_j\} = \theta_i\theta_j + \theta_j\theta_i = 0. \qquad (5.1)$$

An important consequence of this relation is that $\theta_i^2 = 0$. This means that every element of $\mathcal{G}_n$ can be expressed as a sum,

$$f(\theta) = f^{(0)} + f_k^{(1)}\theta_k + f_{k_1k_2}^{(2)}\theta_{k_1}\theta_{k_2} + \ldots + f_{12\ldots n}^{(n)}\theta_1\theta_2\ldots\theta_n, \qquad (5.2)$$

where the highest order monomial is of degree n, and where the summation convention has been assumed.

Berezin also defined left and right derivatives, and integration. For example the right derivative was given by $\frac{\partial \theta_i}{\partial \theta_j} = \delta_i^j$. The rule for products can be obtained by generalizing from:

$$\frac{\partial}{\partial \theta_i}\theta_j\theta_k = \delta_{ij}\theta_k - \theta_j\delta_{ik}. \qquad (5.3)$$

Integration was defined as follows:

$$\int d\theta_i = 0, \qquad (5.4)$$

$$\int d\theta_i\theta_i = 1. \qquad (5.5)$$

An interesting feature of these Berezin integrals is that the operation of integration is like that of differentiation. δ-functions can also be defined,

$$\int d\theta f(\theta)\delta(\theta - \theta') = f(\theta'). \qquad (5.6)$$

Grassmann Algebra in the Dual Models

In 1973 Montonen used Grassmann variables in the study of dual resonance models. The goal of this work was to invent a way to perform multiloop calculations involving fermi operators. It was noted that, in the Veneziano model, general multiloop amplitudes could be constructed by a procedure which involves taking a multiple trace.

The difficult combinatorial problem of calculating the trace could be converted to a more manageable functional integral by introducing coherent states, i.e. eigenstates of the destruction operator. Since such states do not exist for fermi operators, Montonen introduced a "coherent-like state" in terms of anticommuting operators b and Grassmann variables θ.[91]

Soon after this, Fairlie and Martin discovered that it was possible to use Grassmann variables to see that the Neveu-Schwarz model is an extension of the Veneziano model.[37]

The Introduction of Superfields

Salam and Strathdee suggested that the study of supersymmetry might be done most conveniently in a space in which the usual space-time coordinates x_μ are augmented by a set of Grassmann variables θ_α. The most general form for a function of the superspace variables is that given by (5.2) for $f(\theta)$. A function which can be expanded in this way is referred to as a superfield, and will be denoted by Φ. In their first paper on the subject Salam and Strathdee considered Grassmann variables which were Majorana spinors, and they chose to impose a reality condition on the superfield

$$\Phi(x,\theta)^* = \Phi(x,\theta).$$

Salam and Strathdee noted that "the truly remarkable and exciting feature of the Wess-Zumino group is that the superfield $\Phi(x,\theta)$ in eight dimensions is exactly equivalent to a 16-component set of ordinary fields in four dimensions.[105]" The functions of the space-time variables x_μ which multiply each term of the superfield expansion are referred to as the component fields. When the reality requirement is placed on the superfield, the component fields are A, B, ψ, F, G of the scalar multiplet, along with v_μ, λ, D of the vector multiplet. Soon after the paper by Salam and Strathdee, Ferrara, Zumino and Wess showed that it was possible to consider superfields which describe only the scalar multiplet.[41] To do this they introduced two features which have been widely used in work on the superfield formalism ever since. They used two-component spinors, and they constructed superfields (the so-called chiral superfields) which were functions of θ and not of $\bar{\theta}$. A chiral superfield can be expressed in terms of a general superfield which is subject to a constraint. The constraint can be written in terms of a covariant derivative $\bar{D}_{\dot{A}}$. In the two-component notation

$$D_A = \frac{\partial}{\partial \theta^A} - i(\sigma^\mu)_{A\dot{A}}\bar{\theta}^{\dot{A}}\partial_\mu,$$

$$\bar{D}_{\dot{A}} = -\frac{\partial}{\partial \bar{\theta}^{\dot{A}}} + i(\sigma^\mu)_{A\dot{A}}\theta^A \partial_\mu. \tag{5.7}$$

A chiral superfield satisfies

$$D_{\dot{A}}\Phi = 0. \tag{5.8}$$

In Ref. 105 Salam and Strathdee showed that supergauge transformations of the component fields could be derived by considering the transformations of the superfield under the superspace transformations: $\theta \to \theta + \epsilon$, $x_\mu \to x_\mu - \frac{i}{2}\bar{\epsilon}\gamma_\mu\theta$. In particular these transformations imply that the variation of the highest component of the superfield is a derivative, so that its "space-time integral ... would be an invariant if surface effects can be neglected."

The highest component can of course be isolated by performing successive superspace differentiations. However, because of the similarity between superspace differentiation and integration, these invariants can also be expressed as integrals over the superspace variables. Thus, it is possible to express superinvariant Lagrangians and possible counterterms as such integrals. For example the mass and interaction terms of the first interacting theory constructed by Wess and Zumino are of the form:

$$\int d^4x d^2\theta [\alpha\Phi^2 + \beta\Phi^3], \tag{5.9}$$

where Φ satisfies (5.8). Because Φ is chiral, there can be no integral over $\bar{\theta}$.

The elements described above were introduced in the early papers on superfields. They sufficed to develop Feynman diagram methods. References to the early work which developed these methods and demonstrated their usefulness through specific applications can be found in Refs. 108, 109, 40, 132, and in many other sources.

In 1979, Grisaru, Roček, and Siegel proposed a method designed to simplify supergraph calculations by "[making] greater use of supersymmetry."[73] After discussing the advantages that had been associated with the older super-Feynman rules, they noted that "relatively few superfield calculations were carried out... Although simpler than conventional calculations, they were nonetheless cumbersome because the methods did not fully exploit the inherent advantages of supergraphs as a manifestly supersymmetric formalism. It was necessary to manipulate expressions involving explicit spinor variables, and to rely on complicated algebraic identities to obtain simple final results." The improved rules led directly to what have been called 'non-renormalization theorems'. In Ref. 73 itself, Grisaru, Roček, and Siegel proved the theorem that "each term in the effective action can be expressed as an integral over a single $d^4\theta$." The consequences of this sort of result can be appreciated by looking at the superfield form of the mass and interaction terms of the Wess-Zumino multiplet. Since these cannot be expressed as such an integral, no terms of this type can pop up in the effective action, and so no renormalization of the mass and coupling constant is needed.

The superfield techniques were used (for example to show that the β-function for $N = 4$ super-Yang-Mills is zero to three-loop order[74]), and further refined (see for example Ref. 75). Eventually they were used to show that the $N = 4$ super-Yang-Mills theory is finite. (See Refs. 15 and 132 for references.)

The superfield approach was not the only path toward the study of the quantum properties of supersymmetric theories. As a matter of fact, the finiteness of the $N = 4$ Yang-Mills theory was first proved via another path—the use of an 'anomalies' argument. This had been done by Sohnius and West by 1981.[117] There are also references in the literature to an unpublished work by Ferrara and Zumino. Actually the foundation for the anomalies argument was laid in 1974 in a paper by Ferrara and Zumino.[43] In '*Transformation properties of the super-current*', they showed that "the spinor current, if correctly defined, belongs to a supermultiplet, together with the energy-momentum tensor and the axial-vector current." This meant that the anomalies associated with these currents also form a multiplet, so that if some of the associated internal symmetries are preserved, then θ^μ_μ, the trace of the energy-momentum tensor must vanish. However since θ^μ_μ is proportional to the β-function, its vanishing implies that the β-function is also zero, and thus that the theory is finite.

Thus, by 1981, supersymmetry had been understood to have given the world a finite quantum field theory. The progress of hopes for a finite theory of supergravity will be sketched in the next section.

6   Supergravity

6.1   *The Discovery and Development of Supergravity*

Motivation and Early Developments

Quantum gravity is inherently non-renormalizable. This problem, which was noticed by Heisenberg in the late 1930's,[78] arises because the coupling constant κ, which is proportional to the square root of the Newtonian constant G_N, is not dimensionless. This was discussed briefly in Sec. 3, as was the possible string solution suggested by Scherk and Schwarz. There is another way in which the difficulty can be phrased. Because κ has the dimensions of $(mass)^{-1}$, the number of derivatives in possible counterterms is different from the number of derivatives in the original Lagrangian. Therefore the counterterms cannot be reabsorbed through renormalization. This need not be a problem if such counterterms are outlawed by a symmetry principle. A great deal of literature exists on the problem and on possible solutions. (See for example the contributions to Ref. 15.)

It must be stressed that these problems of renormalization are problems of a perturbation expansion, and may not indicate that there are infinities in the non-perturbative predictions of quantum gravity. However most of the work which incorporates supersymmetry into theories of gravity has focused on properties of a quantum perturbation expansion.

In a paper which was published in 1974, 't Hooft and Veltman put the calculation of perturbative effects in quantum gravity on a firm footing.[119] In their paper they calculated all of the one-loop divergences of pure gravity, and of gravity coupled to a scalar field. Although they showed that the divergences that were encountered in their calculation of pure gravity were not physically relevant, there were non-trivial divergences when a scalar field was coupled to gravity. The length and difficulty of the calculations were indicated in the ending to their paper. "We do not feel that this is the last word on this subject, because the situation... is so complicated that we feel less than sure that there is no way out. A certain exhaustion however prevents us from further investigation, for the time being."

This work by 't Hooft and Veltman inspired other (not-yet-exhausted) physicists to explore the situation when gravity is coupled to other matter systems. Gravity coupled to a charged scalar field,[97] to the Maxwell field,[25,26] to spin-$\frac{1}{2}$ fermions,[28] to the Yang-Mills system,[27] and to quantum electrodynamics,[29] were studied. In all cases there turned out to be non-trivial divergences. It is interesting to note that among the most active participants in this search for a finite matter-gravity coupling were Deser and van Nieuwenhuizen. Freedman had also studied gravity-matter couplings, taking a different approach which led to somewhat more hopeful results, by allowing only external gravitons (Ref. 48 and references therein).

Many of the negative results mentioned above had been obtained very soon after the work by 't Hooft and Veltman. By that time tremendous progress had been made toward the goal of a unified theory of the fundamental interactions. The Weinberg-Salam-Glashow model had begun to be experimentally tested, and there was a growing feeling that it might indeed describe a physically realized unification of electromagnetism and the weak force. Grand unification—a unification of the strong, weak, and electromagnetic forces—had been proposed. The force that was still excluded was gravity. Since there was not much hope that gravity could be included unless a consistent and predictive quantum theory of gravity coupled to matter could be formulated, the search for such theory must have seemed particularly urgent.

One idea that was suggested at that time as a possible way to include gravity was that a new symmetry principle should be introduced which might forbid the unwanted counterterms made possible by the dimensionality of κ. It seemed possible that supersymmetry might be just the symmetry to cure the ills of quantum gravity. It had already proved to be capable of softening ultraviolet divergences like those that plagued gravity, and so naturally there was hope that the requirement of supersymmetry would mandate a coupling of the gravitational field to matter for which the ultraviolet divergences would cancel. And there was the further hope that such a coupling would yield a theory in which all of the known forces were naturally unified.

It turned out that even constructing such a supersymmetric theory as a step toward testing these hopes was a Herculean task. The difficulties can begin to be appreciated without even looking at the eventual form of the theories of supergravity. All one needs to do is to note that almost two and one half years elapsed between the first paper of Wess and Zumino and the paper of Freedman, van Nieuwenhuizen, and Ferrara.[51][i]

This time lag was certainly not due to a late-dawning realization of the possibilities of supergravity. In their very first paper on supersymmetry Wess and Zumino had mentioned that it would be natural to consider supersymmetric theories with general coordinate invariance. The reason given there was a formal one, based on the structure of the algebra of transformations. However, once the possibility of canceling divergences in supersymmetric theories had been realized, the potential benefits of combining supersymmetry and gravitation began to be discussed. At the *XVII Conference on High Energy Physics* which took place in London in 1974, Zumino gave a talk on supersymmetry.[116] There he asked, "Can a theory of this kind, because of the compensation of divergences due to supersymmetry, provide a renormalizable description of gravitational interactions?" The possible role of supersymmetry in eliminating the divergences of theories of gravity coupled to matter was later mentioned by others in discussions of the negative results described above. (See for example Deser's contribution to Ref. 116, in which he refers to Zumino's conjecture, and van Nieuwenhuizen's contribution to Ref. 104.)

In other areas a time lag of two or three years might not mean much, but in the fast-moving area of supersymmetry and what was to become supergravity, a great deal would generally happen in two

---

[i] It is generally accepted that supergravity was independently worked out by two groups: D. Freedman *et al.*, Ref. 51, and S. Deser and B. Zumino, Ref. 30, see e.g. P. West, *Introduction to Supersymmetry and Supergravity* (World Scientific, Singapore, 1990). See also footnote on page 233. – Editors' note.

years, and a lot did happen in the time between Wess and Zumino's paper on supergravity. Actually a lot happened in the study of local supersymmetry during those years. Arnowitt, Nath, and Zumino all made contributions to this study, for example Ref. 93 and Ref. 7, that were somewhat in the spirit of the work of Volkov and Soroka on curved superspace.

Gauge Supersymmetry

Richard Arnowitt, Pran Nath, and Bruno Zumino used the superfield formalism to attempt to construct a supersymmetric theory which included gravitation. The basic idea behind this approach was to extend the use of general coordinate invariance to superspace. Thus, the coordinate space that they worked with was the space which included Grassmann coordinates in addition to the four space-time coordinates used in standard general relativity. The general coordinate transformations of Einstein's theory were extended to include general coordinate transformations in superspace. Although Arnowitt, Nath, and Zumino collaborated on at least one paper,[7] there were differences between the approach taken by Nath and Arnowitt, and the one taken by Zumino. These differences had to do with their approach to the geometry of superspace. While Arnowitt and Nath considered a Riemannian geometry of curved superspace, Zumino considered a non-Riemannian geometry. (The differences between the two approaches were discussed at the conference: *Gauge Theories and Modern Fields Theories*, which was held at Northeastern University in the fall of 1975.[8])

The approach of expanding the space of the coordinates was also taken in both the earlier and later work on Kaluza-Klein theories, although there the added coordinates had the same (bosonic) character as the original spatial coordinates. The fermionic character of the extra dimensions of superspace makes them quite different from the space-time coordinates x_μ. Because of this difference it seemed clear that the new coordinates should not be regarded as physical. For the Kaluza-Klein theories, the question of the physical interpretation of the extra dimensions, which are not obviously different from the other spatial dimensions, cannot be ignored. Zumino noted that "the fact that the additional space-time dimensions in the present picture are fermionic and spinorial, instead of bosonic as in Klein-Kaluza, presents the combined advantages that physical space-time is obviously still four-dimensional and that half-integral spin fields arise very naturally.[8]" It is interesting that the addition of either the usual bosonic coordinates, or of fermionic coordinates, leads to theories in which the fundamental forces are unified.

Nath and Arnowitt began their study of gauge supersymmetry, with a paper which was submitted to *Physics Letters* in February of 1975.[93] In their approach, supersymmetry and gravity were combined in an interesting way. The metric of superspace, $g_{AB}(z)$, occupied center stage. The subscripts A and B could denote either indices of the usual four dimensional space-time, or spinor indices. This metric was expanded as a superfield, and the physical fields were taken to be a subset of the components of the superfield.

Nath and Arnowitt noted that, "the usual supersymmetry transformation is ... a linear transformation in the 8-dimensional superspace of $z^A = \{x^\mu, \theta^\alpha\}$. It is natural then to consider a generalization of this to arbitrary coordinate transformations in superspace, $z^{A'} = z^A(z)$ [restricted to maintain the bose (fermi) nature of x^μ (θ^α)]. We take this to be our basic gauge group".[89] They then went on to show that this gauge group includes not only the group of general coordinate transformations of general relativity, but also the other gauge transformations of elementary particle physics, including the Yang-Mills transformations of SU(n).

Furthermore, Nath and Arnowitt showed that the component fields of the metric played the role of gauge fields. They inferred that "if *one extends supersymmetry to be a local gauge invariance then it is possible to construct a theory where all fields are gauge fields*, and hence all interactions are determined by a common non-Abelian gauge invariance."

In comparison to the later theories of supergravity, the conceptually rich gauge supersymmetry did not generate the enthusiasm and active participation of many researchers. It would seem that the reason for this was related to the very richness of the formalism. For, the large number of fields that were automatically included in the metric superfields made the identification of the physics described by the theory difficult. This point is mentioned in the first paragraphs of the work by Freedman, van Nieuwenhuizen, and Ferrara,[51] where they explain the reason for their resolve to work with only physical fields.

The Discovery of Supergravity

Freedman, van Nieuwenhuizen, and Ferrara chose to start with the simplest possible multiplet of physical fields which could be expected to respect a local supersymmetry. They wrote "we commit ourselves from the start to a formulation without superspace in which the only fields in the gravitational supermultiplet are the metric tensor $g_{\mu\nu}(x)$ [or, equivalently, the vierbein field $V_{a\mu}(x)$] and a Rarita-Schwinger field $\psi_\mu(x)$."

If the gravitational field is to couple to a spinor field, as is to be expected in a supersymmetric theory, then it is convenient to use the vierbein formalism. The vierbein is related to the metric by: $g_{\mu\nu} = V_{\mu a} V_{\nu b} \eta^{ab}$. Thus the vierbein can be taken to be the fundamental field, with the variations of the metric derived through its relation to the vierbein, and not considered to be independent. This formalism, in which the vierbein (or, equivalently, the metric) is the only independent field variable associated with gravitation, is referred to as the second order formalism.

The approach taken by Freedman, van Nieuwenhuizen, and Ferrara was to make an ansatz for the form of the Lagrangian and also for the form of the supersymmetry transformations. They then checked to see whether the chosen Lagrangian was symmetric under the chosen transformations. When it turned out that their original choices did not quite work (although they came close), they modified both the Lagrangians and the transformations. In this next step they completed the construction of a theory of supergravity.

The Lagrangian that they started with was the simplest one possible for gravitation coupled to a Rarita-Schwinger field,

$$I = \int d^4x \left[\frac{\kappa^{-2}}{4} \sqrt{-g} R - \frac{1}{2} \epsilon^{\lambda\rho\mu\nu} \bar{\psi}_\lambda(x) \gamma_5 \gamma_\mu D_\nu \psi_\rho(x) \right] \qquad (6.1)$$

The coupling of the spin-$\frac{3}{2}$ field to the gravitational field is just the minimal coupling, and is accomplished through the presence of the covariant derivative in the Lagrangian,

$$D_\nu \psi_\rho(x) = \partial_\nu \psi_\rho(x) - \Gamma^\sigma_{\nu\rho} \psi_\sigma + \frac{1}{2} \omega_{\nu ab} \sigma^{ab} \psi_\rho, \qquad (6.2)$$

where the form of $\Gamma^\sigma_{\nu\rho}$ in terms of the vierbein is the one given by the Christoffel symbol, and where $\omega_{\nu ab}$ represents the so-called spin connection, which, like the metric, can be expressed in terms of the vierbein.

$$\omega_{\nu ab} = \frac{1}{2} \left\{ [V_a^\nu (\partial_\nu V_{b\mu} - \partial_\mu V_{b\nu}) + V_a^\rho V_b^\sigma (\partial_\sigma V_{c\rho}) V_\nu^c] - [a \leftrightarrow b] \right\}.$$
$$(6.3)$$

The ansatz for the supersymmetry transformations was

$$\delta\psi_\mu(x) = \kappa^{-1} D_\mu \epsilon(x),$$
$$\delta V_\mu^a(x) = i\kappa \bar{\epsilon}(x) \gamma^a \psi_\mu(x). \qquad (6.4)$$

The trial expression for the variation of the Rarita-Schwinger fields is a generalization of the symmetry transformation $\delta\psi_\mu = \partial_\mu \epsilon$ of the free Rarita-Schwinger field. The authors noted that this is an attempt "to interpret supersymmetry as the curved-space generalization of the old Rarita-Schwinger gauge transformation." This is analogous to the procedure used by Akulov and Volkov to generalize the symmetry transformations of the Dirac field.

Because the chosen action turned out not to be invariant with respect to these transformations, both the action and the transformations were modified, with a term bilinear in the fermi fields added to the transformations and a corresponding quadratic term added to the Lagrangian. $\delta\psi_\mu$ was augmented by $\frac{i\kappa}{4}(2\bar{\psi}_\mu \gamma_a \psi_b + \bar{\psi}_a \gamma_\mu \psi_b)\sigma^{ab}\epsilon$.

With the action and the supersymmetry transformations so modified, the action was indeed invariant under the local transformations; Freedman, van Nieuwenhuizen, and Ferrara had succeeded in constructing a theory of supergravity. As can be seen from the form of the action and of the transformations, these authors had completed a formidable calculational task. Indeed the complete result required a computer computation.

Another important component of this first paper on supergravity was the study of the algebra of the proposed transformations, to verify that it was indeed a supersymmetry algebra. Of course, because there were no auxiliary fields, closure of the algebra could only be accomplished if the equations of motion were used. The analysis of the algebra was completed in a second paper by Freedman and van Nieuwenhuizen.[52] They found that "the algebra of local supersymmetry transformations, when restricted to fields satisfying the equations of motion, consists of general coordinate transformations plus field-dependent local Lorentz rotations and further field-dependent supersymmetry transformations." This situation is analogous to the situation encountered in other supersymmetric gauge theories, in that in those theories as well, the algebra naturally included field-dependent gauge transformations, as described in an earlier work by Freedman and de Wit.[50]

In discussing the physics associated with their theory, Freedman, van Nieuwenhuizen, and Ferrara speculated that the super-Higgs effect which had been discussed by Volkov and Soroka might allow the massless Rarita-Schwinger field to acquire a mass. This would solve two of the potential phenomenological problems of supersymmetry at once: the mass acquired by the Rarita-Schwinger field would insure that it was not associated with any long range effects which might have been inconsistent with established observations; and the eating of the spin-$\frac{1}{2}$ Goldstone fermion would have eliminated the goldstino. This was especially desirable in light of earlier work of Freedman and de Wit which had indicated that it was highly unlikely that neutrinos, the only known apparently massless fermions, could correspond to the massless Goldstone fermions associated with the spontaneous breaking of supersymmetry.[49]

The construction of supergravity theories suggested or renewed interest in many fields of inquiry. One of the questions that was raised was the consistency of a theory with an interacting spin-$\frac{3}{2}$ field. This question was addressed in a paper which was submitted for publication about a month after[j] the paper by Freedman, van

---

[j] This paragraph as well as the previous discussion of the Freedman *et al.* **vs.** Deser and Zumino issue suggests that Deser and Zumino's work[30] was secondary with regards to that of Freedman *et al.*[51] The editors believe that it would be more appropriate to consider both works, Refs. 30 and 51, on equal footing. They present independent constructions of supergravity, as is generally accepted in the world literature. – Editors' note; see also footnote on page 228.

Nieuwenhuizen, and Ferrara. This paper, which was written by Deser and Zumino, also intrinsically raised another more technical question. This question was, what is the best way to approach the study of a theory of gravity which requires long and intricate calculations? That is, is there a formalism which might make such calculations more tractable? The paper of Deser and Zumino will be reviewed below during the discussion of the development of supergravity.

The Study of Supergravity

Immediately after the construction of the first theory of supergravity, the issues that needed to be clarified in order to smooth the way for further progress were addressed. Because the derivation of results about supergravity presented substantial technical difficulties, it was necessary and not at all trivial to think through questions about the most appropriate formalism. For example: what was the best way to handle the gravitational field in a theory which necessarily includes spinors? How best to approach the development of the superfield formalism? Was the component formalism superior to the superfield formalism or vice versa? There was also the question of whether fields of spin even higher than 2 could be consistently included. And perhaps most important, given the primary reason for the development of supergravity in the first place: could theories of supergravity be well defined at the quantum level?

There were other important questions as well. Perhaps most central to understanding whether or not theories of supergravity could be physical theories was the study of symmetry breaking and the construction of phenomenological models. Because of limits of time and space, these other issues will receive only glancing attention in this presentation. However, one can gain insight into the type and pace of developments in these areas, as well as in the areas which will be somewhat more extensively covered here, from examining the proceedings of the many conferences and schools that have been held on supersymmetry and supergravity. The schools which were held at Trieste in 1981,[45] 1982,[46] and 1984,[33] are particularly useful in this regard, especially concerning the increasing use of formulations in higher dimensional space-times. These will be discussed in Sec. 7.

The conference that was held at Stony Brook in 1979 provides some insight as to the state of affairs when supergravity was about two and one half years old.[122] The study of supergravity in superspace was well underway. Almost one third of all of the papers that were included in the proceedings were about the progress that had been made toward understanding how best to explore supergravity in superspace (including the derivation of superspace constraints), or about results that had been obtained through use of the superfield formalism, or dealt with a comparison between the superfield and component field approaches. Another fairly large group of papers (about 13% of the total), was devoted to the study of coupling among fields which include fields of higher spin. Papers that dealt with the problem of supersymmetry breaking and of the phenomenology of theories of supergravity comprised about 15% of the total. Naturally a significant number of papers (not necessarily distinct from the papers on superfields) dealt with the study of quantum supersymmetry. There were also a number of papers (over 10%) which dealt with the topological features of supergravity theories. It is clear that a lot of creative thought about supergravity had been going on, for there were a number of papers with provocative titles. For example: '*Local Supersymmetry without Gravity?*' (Townsend), '*From Supergravity to Antigravity*' (Scherk), '*Difficulties in Constructing Bose Analogs to Supergravity*' (Milton and Urrutia), and '*Complex and Quaternionic Supergeometry*' (Lukierski).

The Consistency of Theories Which Include Fields with Spin Greater than One

Long before the invention of supersymmetry, studies of interacting spin-$\frac{3}{2}$ fields had encountered problems of consistency. Since the requirement of supersymmetry demanded the existence of at least either one spin-$\frac{3}{2}$ field or one spin-$\frac{5}{2}$ field as a partner for the graviton, the consistency of theories of supergravity required that at least those interactions mandated by the requirement of local supersymmetry be consistent.

This question was first addressed in the paper by Deser and Zumino on supergravity.[30] As a matter of fact it was this consideration that led them to their form for the variation of the fields. They observed that in the theories for which the coupling of the Rarita-Schwinger fields to other fields had been shown to lead to inconsistencies, the requirement of consistency led to too restrictive constraints on the space of solutions to the equations of motion. They discovered that the requirement of local supersymmetry ensured that these would-be constraint equations were automatically satisfied when the equations of motion were, thereby eliminating the problem.

Although the consistency of the (spin-2)-(spin-$\frac{3}{2}$) interaction had been ensured through the requirement of local supersymmetry, it remained to discover whether or not a consistent coupling to spin-$\frac{5}{2}$ fields was also possible. Supergauge theories which included these higher spin fields were referred to as theories of hypergravity. In the years just after the discovery of supergravity, this was an active area of research. However the results were discouraging, and it came to be widely felt that, because of problems with interactions, fields with spin greater than two could not be consistently incorporated into theories of supergravity. This had an important consequence, because it placed a limit on the amount of internal symmetry which a theory of supergravity could accommodate. If the highest spin allowed in the multiplet was to be equal to two, group-theoretic considerations limited the number of supersymmetry generators, N, to 8. As will be discussed below, this made it seem less likely that theories of supergravity could be shown to give finite quantum results.

Questions of Formalism

Although this paper deals primarily with the conceptual development of supersymmetry, the complexity and flavor of some of the calculations which led to a greater understanding of the subject is such that the conceptual development cannot be fully appreciated without making contact with the formalism. This was true for the superfield techniques which have played such a crucial role in the construction of supersymmetric theories and in the study of their quantum properties. It is also true of supergravity.

Second order, first order, and 1.5 order formalism: The theory of gravity is difficult to work with, and the ability to understand its quantum predictions can be enhanced by choosing or inventing a formalism that is well suited to the study of the question at hand. When the gravitational action is expressed in terms of the vierbein, its functional form is quite complicated. A substantially simpler form is achieved by introducing the spin connection. When one goes a step further, and considers $\omega_{\mu ab}$ as a field which can be varied independently of the vierbein, then the work is said to be done in the first order formalism. Deser and Zumino chose to work in the first order formalism.[30]

The physics discussed in the paper of Deser and Zumino was the same as that in the paper of Freedman, van Nieuwenhuizen, and Ferrara. However, in the first order formalism the functional form of the spin connection in terms of the vierbein and the other fields of the theory, is given by the Euler-Lagrange equations. It turned out that the expression that was so derived by Deser and Zumino was different from that given by (6.3). Nevertheless, when their expression for $\omega_{\mu ab}$ was substituted back into the Lagrangian, the complete Lagrangian of Ref. 51 was obtained. Thus, even those interaction terms which were quadratic in the Rarita-Schwinger field, the terms that had to be added in the second step of the calculations in Ref. 51, were automatically included.

It turned out to be possible to simplify the calculation of the variation of the Lagrangian still further, by using a formalism which has been called the 1.5 order formalism. This simplification is based on the fact that, because of the form of the equation of motion associated with the spin connection, $(\frac{\delta \mathcal{L}}{\delta \omega_{\mu ab}} = 0)$, it is possible to neglect those terms in $\delta \mathcal{L}$ which are proportional to the variation of the spin connection.[123]

The importance of an appropriate formalism has recently been underscored by the introduction of a set of new variables to describe Einstein's gravity. These new variables, which were introduced by Abhay Ashtekar, seem to be well suited to the study of quantum gravity. (The reason for this has to do with the simplification of the equations of constraint which occurs when the new variables are used.) Ashtekar's formalism has already been applied to supergravity by Jacobsen. (See Ref. 9 and references therein.) However, the extent of the impact that the new variables may eventually have on the study of supergravity, is not yet known.

Supergravity in superspace: Another concern that needed to be addressed was the question of the superspace formulation of supergravity. In order to construct a theory whose physical content was clear, it had been necessary to leave superspace. This journey had necessitated giving up the advantages that had been associated with superspace. These include: a set of linear transformations whose algebra is closed without the need to invoke the equations of motion, a systematic method for the construction of further theories, and simplified quantum calculations.

A natural step to take after the successful construction of supergravity, was to try to find a set of auxiliary fields to complement the minimal set of dynamical fields used in that construction. This step was taken by Ferrara and van Nieuwenhuizen,[44] and by Stelle and West.[118] This work was followed by a great deal of further work on the study of supergravity in superspace. (For references see Refs. 57 and 77.) As is indicated by the large number of contributions devoted to this subject at the Stony Brook conference, this work and its comparison to work without superspace attracted a great deal of attention in the years just after the discovery of supergravity. Eventually it would prove possible to study even the promising $N = 8$ supergravity theory in the superfield formalism. This study would, by 1981, lead many experts to doubt that even this most symmetric supergravity theory could be shown to yield finite quantum results.

Supergravity in higher dimensions: A third question of formalism arose with the use of higher dimensional space-times to study theories of supergravity. This will be the focus of Section 7.

The Study of Quantum Supergravity

Since one of the primary motivations for the introduction of supergravity was the possibility of discovering a gravity-matter coupling which led to a well-defined quantum theory, the study of quantum supergravity began immediately after the first construction of a theory of supergravity. Two approaches to this work were taken. One concentrated on the study of the S-matrix, and the other concentrated on the study of possible counterterms. The fact that those results which were arrived at both ways agreed, gave confidence in the validity of the results.

Grisaru, van Nieuwenhuizen, and Vermaseren computed the one-loop corrections to photon-photon scattering and found that the divergences canceled against each other.[70] In the same paper they examined pure supergravity and found it to be one-loop finite. Grisaru went on to use S-matrix-oriented techniques to show that pure supergravity is also two-loop finite.[72] (See Ref. 71 for a discussion of the S-matrix approach.)

The philosophy behind the search for possible counterterms was that, if such terms could be shown to be impossible, then one would know that there were no problems with non-renormalizability. However, if such terms were found to exist, then their very existence is a signal of the possibility of trouble for the quantum theory. Although it would not necessarily mean that there are problems—since the coefficient of such terms in the quantum perturbation expansion could turn out to be zero, especially it there are other symmetries at work—it would nevertheless be a discouraging sign. Deser, Kay, and Stelle found that there were possible invariants for pure supergravity from three loops onward.[31]

Since it was conceivable that other symmetries would constrain the coefficients of these bothersome counterterms to be zero, it seemed advantageous to study supergravity theories which had more symmetry, i.e. the theories with values of $N > 1$. However, Deser and Kay found a possible three loop counterterm for $N = 2$ supergravity.[32]

Using superfields methods, Howe and Lindström studied the theory with maximal symmetry, the $N = 8$ theory, and found that the theory seemed to have non-vanishing invariants from seven loops onward. (See their contribution to Ref. 77.)

Although they indicated that it might be possible to circumvent theory result, they concluded that "On the face of it, it would therefore seem as if supergravity is only an improvement on quantum gravity in the sense that the divergences start at higher loop order."

This result was presented at the Nuffield workshop on superspace and supergravity, which was held in the early summer of 1980.[77] It seems to have brought the curtain down on most hopes that supergravity could itself cure the ultraviolet problems of quantum gravity coupled to matter. In his *Physics Reports* review on the situation,[123] Peter van Nieuwenhuizen recapped the results on quantum supergravity, and commented that the "results seem to leave little hope short of a miracle ... Of course, miracles do sometimes happen, but they would have to happen at every loop order. It seems better to accept a fundamental property of gravitons: that no nearby massless theory exist (the van Dam-Veltman theorem) and to use nonperturbative methods. Perhaps this is the way quantum supergravity should go in the future, but it is easier to say that one should do nonperturbative calculations than to do them."

These discouraging results set the stage for the next act, in which the possible role of strings in eliminating the ultraviolet divergences was (and is being) explored.

7   Supersymmetry in Higher Dimensions

A great deal of work in supersymmetry has been done in the context of space-times that have more than four dimensions.

Higher dimensional space-times had been used in the study of gravitation before as a possible means of unifying gravitation with the other forces. Kaluza's original idea was written up in 1919 (although it was not published until 1921), and so applied only to gravitation and electromagnetism. This idea was also studied extensively by Klein and many others. (A history of Kaluza-Klein theories has been given by Appelquist, Chodos, and Freund.[6] Those references which are not specifically cited below may be found there.) Although there had not been a great deal of activity centered around the Kaluza-Klein approach to unification in the recent past, Bruce De Witt assigned the problem of unifying gravitation and Yang-Mills forces in the Kaluza-Klein geometrical framework as a problem (no. 77) in his 1963 Les Houches lectures. The effect that this assignment had on the students who attended the lectures and on the many more who read the lecture notes is difficult to gauge. Nevertheless, by 1968 Kerner had presented work on this project. Cho and Freund later added to this, and another important piece of work, which will be discussed further below, was done by Cremmer and Scherk.[19]

Given the past link between the Kaluza-Klein approach to gravitation and the possibility of unification, it was perhaps inevitable that higher dimensions would play an important role in the study of theories of supergravity. There was, however, another road that had led to the study of theories in higher dimensions, and this was the road marked out by the study of dual string theories.

It was primarily through the work of theorists who came to the study of higher dimensional theories via work on hadronic stings, that supersymmetry first came to be studied in higher dimensions. For example the work by Cremmer and Scherk on spontaneous compactification was motivated by considerations that came out of the study of dual models. It is therefore necessary to look at some of this earlier work, in which the role of supersymmetry was largely irrelevant, in order to get a sense of how and why supersymmetric theories first came to be formulated in space-times of more than four dimensions.

7.1   Higher Dimensions of Dual Models

One of the problems of the dual resonance model was the presence of ghost states. A great deal of study was devoted to understanding how these non-physical states could be made to decouple from the physical states. One suggestion in this regard had been to add a fifth dimension. Some work in a 1970 paper by Lovelace indicated that the problem might be solved if the dimensionality of the space-time was increased to 26. This was verified by later work, which also indicated that for the models that were associated with graded algebras, the critical dimension was 10. (References can be found in Ref. 69.)

The existence of a critical dimension seems to have been viewed as something of a mixed curse. On the one hand, since the world does not appear to be 26 (or even 10) dimensional in any obvious way, the known consistent string theories seemed to be unphysical. It seems to have been assumed by most that when a more realistic theory could be constructed, the number of space-time dimensions predicted by it would be four. The positive aspect was that there *was* any prediction of the number of space-time dimensions at all.

Even general relativity, which describes the geometry of space-time, had always been done with the number of dimensions as input.

Although the prevalent attitude seemed to be that the extra dimensions did not themselves have any physical significance, there were individuals who explored the possibility that the extra dimensions might indeed have a physical interpretation. Notable among these were Scherk, Schwarz, and Cremmer.[111,18] This view of the possible physical interpretation of the extra dimensions was part of the larger picture in which the possible role of the dual models had been expanded to include, in addition to the hadronic interactions, at least gravitation as well. Although the themes of string theories as potential 'theories of everything' and the role of the higher dimensionality of string theories are intimately related, in this section we will focus on the aspect of higher dimensions.

7.2   The Physical Interpretation of Extra Compact Spatial Dimensions: an Extrapolation in the Context of Dual Models

In a paper which was submitted to *Physics Letters* in March of 1975,[112] Scherk and Schwarz asserted that "it is possible to make sense of a theory formulated in $4 + N$ dimensional space-time, provided that the extra N variables span a compact and spacelike N-dimensional domain. ... The existence of the N extra spatial dimensions is unobservable so long as one probes with energies less than $1/R$ [where the size of the compact space is of order R]. This is a simple consequence of the fact that fields with non-zero values for the quantum numbers associated with the eigenfunctions of [the compact space] have mass of order $1/R$."

This theme was explored in more detail in a paper by Cremmer and Scherk which appeared soon after this.[18] In '*Dual Models in Four Dimensions with Internal Symmetry*', Cremmer and Scherk studied the implications of interpreting the extra dimensions of the dual theories as spatial dimensions of a compact space. They found that, if the total number of dimensions was equal to the critical number, then there are still no ghosts. Although their central interest was in showing that dual models defined on such spaces are consistent, Cremmer and Scherk included a long and illuminating discussion on field theories in which some of the dimensions are unbounded and some are compact. Their approach included a number of features that were important to the later work that was done on higher dimensional supersymmetric theories.

The particular space that they considered was a four-dimensional space, with two unbounded dimensions and two compact dimensions. x^0 and x^1 corresponded to the unbounded time and space coordinates respectively. x^2 and x^3 will be taken to represent the coordinates of the compact internal space, which was chosen to be a torus of radii R_2 and R_3. Thus, fields with x^2 and x^3 dependence are necessarily periodic. Any field $\Phi(x^0, x^1, x^2, x^3)$ can therefore be written as

$$\Phi[\hat{x}] = \frac{1}{\sqrt{R_2 R_3}} \sum_{n,m} \Phi_{n,m}(x) \exp\left[2i\pi\left(\frac{nx_2}{R_2} + \frac{mx_3}{R_3}\right)\right], \qquad (7.1)$$

where n and m are integers which run from 0 to ∞; hatted variables and indices denote four dimensional quantities.

The authors choose an action:

$$S = \int d^2x \int_0^{R_2} dx^2 \int_0^{R_3} dx^3 (\frac{1}{2}\eta^{\hat{\mu},\hat{\nu}}\partial_{\hat{\mu}}\Phi\partial_{\hat{\nu}}\Phi - \frac{1}{2}\mu^2\Phi^2 - \frac{1}{4!}\lambda_0\Phi^4). \quad (7.2)$$

When Φ is substituted into the action, and the integrals over x^2 and x^3 are performed, the result is an action with kinetic energy and mass terms given by

$$S_{ke} + S_m = \int d^2x (\sum_{n,m} \frac{1}{2}\partial_\mu\Phi^*_{n,m}\partial_\mu\Phi_{n,m} - \frac{1}{2}\mu^2_{n,m}\Phi^*_{n,m}\Phi_{n,m}). \quad (7.3)$$

The mass parameter $\mu^2_{n,m}$ can be expressed as

$$\mu^2_{n,m} = \mu_0^2 + \left(\frac{n^2}{R_2^2} + \frac{m^2}{R_3^2}\right). \qquad (7.4)$$

Cremmer and Scherk went on to interpret the physical meaning of this: "Suppose we would live in such a world. ... Then in high energy experiments one would produce heavy particles in pairs, having two additive conserved quantum numbers n and m... A particle moving in the x_2, x_3 direction would not at all disappear from the two dimensional space-time, but would simply be seen as a particle with n, m not zero. So in this universe we would have no classical feeling at all that there are four dimensions..."

The ideas presented in this paper were important in two ways. Of course, first of all, the idea of how the existence of extra physical dimensions might be exhibited through the existence of high mass particles was important. But, in addition to this, it turned out that the method of dimensional reduction presented there was very useful for supersymmetric theories. The reason for this is that it provides a mechanism for the breaking of supersymmetry. One can get a sense that this might be so from the fact that the procedure outlined above changes the masses of the particles associated with the fields of the theory. Since the masses of all of the particles in a supermultiplet are the same, if the mechanism can change the masses of particles in the same multiplet by different amounts, then the supersymmetry would no longer be manifest.

7.3   Precedents to This Early Work

The paper by Scherk and Schwarz referenced the earlier Kaluza-Klein work which had been done in the context of gravitation. In addition to noting the work of Kaluza, Scherk and Schwarz referenced a work by Thirring which had just appeared in 1972; this particular paper was also referenced in a good deal of the early work on supersymmetric Kaluza-Klein theories.

Work by Ne'eman, and by Ne'eman and Rosen was also cited. This work was not done in the context of gravitation, but was entirely motivated by considerations central to particle physics. However the idea put forward was related to the Kaluza-Klein idea, in that the suggestion made was to view the internal symmetries as being related to transformations of additional internal compact space-time coordinates.

Scherk and Schwarz also mentioned that when Fubini and Veneziano had introduced a fifth dimension to try to deal with the problem of ghost states in the Veneziano model, they had taken the momenta associated with the extra spatial dimension to be discrete.

7.4 Spontaneous Compactification in Kaluza-Klein Theories

Although the Kaluza-Klein idea had been applied to Yang-Mills theories, there was a problem associated with this application that had not occurred when electromagnetism alone had been linked to gravitation. The problem had to do with the identification of the vacuum state. It was felt that the vacuum state should be a direct product of four-dimensional Minkowski space and an $(N-4)$-dimensional compact space. This form for the vacuum solution could not be achieved with a pure Kaluza-Klein theory in more than five dimensions. [By a 'pure' Kaluza-Klein theory is meant one in which the only field considered in the higher dimensional space is the gravitational field.] And, in order to incorporate interactions which went beyond electromagnetism, more than five dimensions were needed.

In 1976 Cremmer and Scherk demonstrated that a product vacuum solution of the desired form could exist if fields other than gravity were included in the higher dimensional space itself.[19] The solution that was exhibited by Scherk and Cremmer was an example of a 'spontaneous compactification'. The compactification is spontaneous in the same way that the breaking of a spontaneously broken internal symmetry is spontaneous. That is, the action and the equations of motion are completely symmetric (in the Kaluza-Klein case they are symmetric in all of the space-time coordinates), but the ground state solution necessarily breaks the symmetry.

Although the inclusion of fields other than the gravitational field in the higher dimensional space seemed to be a violation of the spirit of the Kaluza-Klein approach, it turned out to be ideally suited to the study of supergravity theories, since the existence of fields other than the gravitational field is mandated by the requirement of supersymmetry anyway.

7.5 Supersymmetry in Higher Dimensions

It turned out to be advantageous to study even theories with only global supersymmetry in higher dimensions. One reason for this was that it was possible to derive supersymmetric theories in four dimensions from theories originally formulated in higher dimensions, by a dimensional reduction procedure. As a matter of fact, theories with sufficient complexity to be thought of as serious candidates for physical theories with unified interactions were originally derived in higher dimensions. This happened for both supersymmetric Yang-Mills theories and for supergravity.

The number of independent fields in the higher dimensional space is smaller than the number of independent fields in the lower dimensional space, and so there are fewer fields and indices to keep track of. It is therefore often easier to construct and to work with the theory in its higher dimensional form. For example, through dimensional reduction to four space-time dimensions, the 10-dimensional $N = 1$ Yang-Mills theory is transformed into a Yang-Mills theory with $N = 4$, while the $N = 8$ supergravity theory in 4 dimensions was derived from the $N = 1$ theory in 11 dimensions.

Just as it proved to be easier to derive some supersymmetric theories in higher dimensions, it also proved to be easier to explore some of their properties in these higher dimensional space-times. Four issues which were important in the study of supersymmetric theories are discussed in the subsections just below. There are other interesting issues however, and a vast literature exists. References can be found in many places, for example see Refs. 6 and 23.

Is 11 the Maximum Number of Dimensions?

The 11-dimensional theory from which the maximally symmetric, $N = 8$, supergravity was derived is of particular importance. This is so because, as was shown by Nahm in 1978, any theory of supergravity which is formulated in more than 11 dimensions will, upon dimensional reduction to 4 dimensions, necessarily have N greater than 8, and will, therefore, include fields with spin greater than 2.[139] Since, despite a great deal of study, it seemed as if such theories were inconsistent, this was taken to mean that 11 was the maximum number of dimensions appropriate for the study of supergravity. Furthermore, because the 11-dimensional theory corresponds to the maximally symmetric theory, it was also considered to have the best chance of ultimately being shown to be finite. 11 was a favored number of dimensions for another reason as well, because in 11 dimensions there are no matter multiplets, i.e. multiplets which do not include gravitation. Thus, the requirement that there be just one graviton uniquely specifies the multiplet.

Spontaneous Compactification of 11-Dimensional Supergravity: 11=4+7=7+4

In a paper which was written in 1980, Freund and Rubin commented on the need for additional bosonic degrees of freedom in order for spontaneous compactification to work: "Such a complication of the original theory is contrary to the spirit of unification. Here we remark that bosonic fields other than gravity, automatically appear in higher-dimensional supergravity theories, and that they can induce spontaneous compactification. The number of compactified dimensions can then take only two preferential values. For 11-dimensional supergravity we find that either 7 or 4 space-like dimensions compactify. In the first case ordinary "large" space-time has therefore 1 time and 3 space dimensions, a pleasing result."

Supersymmetry played a key role in the derivation of this result. The total number of dimensions, d, was assumed to be 11, because of the considerations mentioned in Sec. 7.5 above. One of these dimensions was taken to correspond to a time coordinate, and the remaining 10 dimensions were assumed to be spatial. The additional bose fields needed for the compactification were associated with an antisymmetric tensor of rank $s-1$, with corresponding field strength $F^{\mu_1\cdots\mu_s}$ of rank s. Without the requirement of supersymmetry, s is an arbitrary parameter. However the requirement of supersymmetry, specifically the associated equality of the number of fermi and bose degrees of freedom, mandates the existence of a field with rank $s = 4$. Freund and Rubin assumed the existence of a vacuum solution of the product form, and then derived expressions for the scalar curvatures in each of the factor spaces in terms of d and s,

$$R_{d-s} = \frac{(s-1)(d-s)}{d-2}\lambda, \qquad R_s = -\frac{s(d-s-1)}{d-2}\lambda. \qquad (7.5)$$

For $d = 11$ and $s = 4$, the signs of these expressions differ. This indicates that either the four-dimensional space-time is not compact, and the 7-dimensional space is, or a 7-dimensional space-time is not compact, and a 4-dimensional spaces is. Unfortunately, it proved possible to construct an example of the latter type of split,[102] so that it could not be argued that supersymmetry unambiguously leads to a prediction of our observed 4-dimensional space-time.

Higher Dimensions for the Spontaneous Breaking of Supersymmetry

Another way in which the extra dimensions turned out to be useful was in the breaking of supersymmetry. Scherk and Schwarz and Cremmer used an idea similar to that outlined above from the paper of Scherk and Cremmer, to devise a way to affect supersymmetry breaking through dimensional reduction. (References can be found in Schwarz's contribution to Ref. 122.) The phrase 'dimensional reduction' can be used in several ways. In their paper titled '*How to Get Masses From Extra Dimensions*', Scherk and Schwarz note that: "In 'ordinary' dimensional reduction, the coordinates of a $D + E$ dimensional theory are divided into D space-time coordinates (x^μ) and E internal coordinates (y^α) that form a compact space. The fields and symmetry transformation laws that the theory possesses in $D + E$ dimensions are taken to be y independent. ... [A] limitation of this approach is that starting from a massless theory (as is necessarily the case for supersymmetry theories in 10 or 11 dimensions) the resulting reduced theory is also massless (aside from possible masses due to the Higgs mechanism or topological excitations). Also it is exactly invariant under the supersymmetry algebra, implying Fermi-Bose degeneracy that needs to be broken in order to describe the real world." They go on to discuss their generalization of this approach, "The new ingredient is to allow the fields and transformation laws to depend upon the internal y coordinates in a well-defined fashion determined by a symmetry in $D + E$ dimensions."

The fields are expressed in a form that automatically respects the chosen symmetry, and then this form is substituted into the action. Work by Scherk, Schwarz, and Cremmer showed that it is possible, by following this procedure to induce a spontaneous breaking of supersymmetry. In his contribution to the 1979 conference at Stony Brook, Schwarz illustrated how this sort of supersymmetry breaking works. He gave two examples. In each, one starts with an action that has two or more symmetries, two of which are played off against each other. In his first example, he considered an action which describes only bosonic states. The action is like (4.12), but without the mass and interaction terms. The theory is defined on a 5-dimensional space-time, with one of the coordinates, y, assumed to parameterized a circle. This abbreviated action is symmetric under both global phase transformations ($\Phi \to e^{i\alpha}\Phi$) and constant translations ($\Phi \to \Phi + c$). The former symmetry is used to write a specific form of a solution which is periodic in y,

$$\Phi = e^{imy} \sum_{-\infty}^{\infty} \Phi_n(x) e^{\frac{2\pi i n y}{L}}. \tag{7.6}$$

When, in a procedure which parallels the one followed by Cremmer and Scherk,[18] this form is substituted into the action, mass terms are derived, with the mass associated with Φ_n given by:

$$m_n = |m + \frac{2\pi n}{L}|. \tag{7.7}$$

The mass terms in the action indicate that the translational symmetry has been lost. This result holds even after the $L \to 0$ limit is taken and the dimensional reduction is complete.

The procedure for the spontaneous breaking of supersymmetry in a theory of supergravity is similar. The major complication is that fermions need to be included, and that the dimensional reduction will therefore necessarily involve a dimensional reduction of the appropriate Dirac matrices as well. (Actually Cremmer and Scherk had already considered fermions in their paper of 1975.)

In order to illustrate how things work out for a theory of supergravity, Schwarz considered a 4-dimensional theory, whose symmetries include both local supersymmetry and a global chiral symmetry: $\psi_\mu \to (e^{im\Gamma_s})\psi_\mu$, where the indices are considered to be 4-dimensional. This chiral symmetry was used to stimulate a spontaneous breaking of the local supersymmetry which allowed the gravitino to become massive (while the graviton remained massless), when the theory was reduced to a three dimensional one.

Evolving Attitudes Toward Work in Higher Dimensions

One can study the evolving attitude toward work in higher dimensions by looking back at the proceedings of the schools and conferences that have been held on supersymmetry. Higher dimensional theories were discussed at the Stony Brook conference in 1979,[122] but the work on higher dimensions formed a small percentage of the work that was presented there. It is also significant that, at least in their written form, the papers that did use higher dimensions, did not discuss the physical significance of the extra dimensions. Rather, the extra dimensions seem to be included only as a means of learning something about supersymmetric field theories in four dimensions. For example, the contribution of John Schwarz which discussed supersymmetry breaking through dimensional reduction is entitled '*How to Break Supersymmetry*'. There was a contribution by Ulf Lindström, describing work which had been done in collaboration with Deser, entitled '*Use of Dimensional Reduction in the Search for Supergravity Invariants*'. In explaining their motivation for working with a higher dimensional theory, Lindström wrote: "Since the simple supergravity action in 11D is far easier to handle than that of the reduced theory in 4D, the search for a current multiplet (generalizing the Bel-Robinson tensor) and its corresponding invariant should be considerably easier." A paper by D'Adda, A'Auria, Fré, and Regge discussed work that was to become important for the study of supergravity in higher dimensions. Although the paper, '*Geometrical Formulation of Supergravity as a Theory on a Supergroup Manifold*', focused on four-dimensional theories, the authors noted that the question of how their formalism worked in any number of dimensions had already been addressed.

The proceedings of the schools on supersymmetry which were held at Trieste provide an opportunity to see which ideas had made it into the mainstream of work on supersymmetry. The topics presented there were chosen so as to educate graduate students and others who were new to the study of supersymmetry, in what were considered to be its fundamentals, and to guide them toward a working knowledge of the basic tools of the trade. Each of these volumes contains an introduction by Salam which provides an additional guide. Studying the topics of the lectures, one does find an increase in emphasis on Kaluza-Klein ideas. In the proceedings of the 1981 school, there is very little mention of higher dimensions, except in the context of dimensional reduction, particularly the derivation of $N = 8$ supergravity in four dimensions from $N = 1$ theory in 11 dimensions. This is also the only context in which Salam mentioned higher dimensions in his introduction. By 1982 things were different. Salam wrote: "The new ground (relative to 1981) lies in the Kaluza-Klein domain and its compactification discussed by J. Strathdee and M. Duff, with related discussions by H. Nicolai on $N = 8$ supergravity, and a contribution from P. Fré on supergravity in 11 dimensions in the context of Cartan integrable systems." The emphasis on Kaluza-Klein theories was even more pronounced in the proceedings of the 1984 school. Salam wrote, "Far from being a peripheral interest, these theories have now come to occupy the center of the stage among supergravity models. ... This shift has been discernible ever since 1979, when Cremmer, Julia and Scherk showed that the maximal $N = 8$ supergravity massless particles in four dimensions could be obtained as a truncation of the $N = 1$ supergravity theory in eleven space-time dimensions. This was the extended super-Kaluza-Klein miracle. The change since 1979 is the deeper study of the compactification mechanism from the higher to the four-dimensional anti-de Sitter and, in some cases, Minkowski space-time, an appreciation of their topological significance, the spectra of the massive (super)-Kaluza-Klein excitations, investigations of their stability properties as well as of the possible supersymmetry breaking. During the Workshop were considered other aspects of the Kaluza-Klein theories like the problems of inducing the compactifications radiatively, the possible relevance of (super)-Kaluza-Klein theories to phenomenology and cosmology— and, among the other riches, the super-string models ... which after all provided of the original reasons for the higher (ten-) dimensional theories."

Is 11 the Minimum Number of space-time Dimensions?

One of the papers which was responsible for the increased interest in the Kaluza-Klein theories which was evident by 1982, is a paper by Witten,[134] *'Search for a Realistic Kaluza-Klein Theory'*. In this paper Witten made an observation that gave hope that the 11 dimensional supergravity theory might be chosen by phenomenological considerations. He argued that 11 was the lower bound on the number of space-time dimensions, since it is the minimum dimension of a compact space which can have an $SU(3) \times SU(2) \times U(1)$ symmetry. He commented that, "It is certainly a very intriguing numerical coincidence that [the maximum number of dimensions] for supergravity, is the minimum number in which one can obtain $SU(3) \times SU(2) \times U(1)$ gauge fields by the Kaluza-Klein procedure. This coincidence suggests that the approach is worth serious consideration." This paper was submitted to *Nuclear Physics* in January of 1981. The approach it advocated received a great deal of attention during the following three or so years.

Chiral Fermions

There was however a problem with 11 dimensions, which was noticed in Witten's paper itself. That problem had to do with the possibility of constructing a theory in which the non-compact 4-dimensional world would be inhabited by the chiral fermions that are observed in our own apparently 4-dimensional world. In considerations which were studied in both Ref. 134 and in later work, Witten found indications that it might be impossible to construct such a theory by starting in 11 dimensions. This problem also received a great deal of attention from others and several possible loopholes were found. (For a review of these see Ref. 6.) Work by Wetterich,[133] indicated that the problem with chiral fermions could be solved in 2 mod 8 dimensions.

Back to the Higher Dimensions of String Theories

The considerations above provide evidence of a sort of positive feedback loop between string theories and supersymmetric field theories. For, the study of supersymmetry in higher dimensions, which was often initiated by researchers who had worked with hadronic strings, eventually led to a more ready acceptance of the higher dimensions which appeared to be necessary for a consistent quantum supersymmetric field theories. As mentioned above, considerations of chirality in supersymmetric filed theories even led to an independent argument that there might be something special about 10 dimensions, the critical number of dimensions for supersymmetric string theories.

The theme of mutual reinforcement between supersymmetric string and field theory will also be sounded in the discussion of superstrings in the next section.

8 Supersymmetry and Superstrings: Back to the Future

As a graph of work on dual theories [Fig. 3] demonstrates, interest in these theories never completely dried up. However most of the papers that were published about strings in the late 1970's and in the first years of the 1980's were still not about strings as possible fundamental theories. Instead, they considered the role that strings might play as effective theories of the strong interactions. There were some notable exceptions however, and the fruit that these bore was the modern superstring. Supersymmetry played an important role in this development. These notes will outline that role, but will necessarily stop at about 1984.

When we spoke of strings in Sec. 3, Scherk and others had begun to explore the relation of string theories to local field theories. As discussed in that section, the understanding that it was possible to obtain both Yang-Mills theories and gravitation as low energy limits of string theories was largely responsible for the proposal by Scherk and Schwarz to view string theories as possible theories of everything.

It would seem that, beyond its conceptual significance, the implication that string theories contain local field theories, was important to the eventual development of superstrings in another more practical way. For it meant that those who still maintained a strong interest in exploring string theory did not have to part ways with their fellow theorists who were primarily interested in studying fields, since studies of field theory could yield information that would be important to the study of strings, and vice versa. In fact, the study of supersymmetric field theories did provide valuable leads for the investigation of strings, while taking the low energy limit of certain string theories led to the derivation of new supersymmetric field theories. The possibility of such a symbiotic relationship was of course made more likely by the string enthusiasts, like Scherk and Schwarz, who also worked on supersymmetric field theories. For it allowed them to help to set an agenda for the study of supersymmetry in which some of the ideas from string theory became a familiar part of the repertoire of many physicists whose interest was primarily in field theory. When the work of Green, Schwarz, and others indicated the there were important reasons to study strings, the necessary tools were within the reach of a relatively large number of active physicists, and some of the ideas (like the possible need to work in higher dimensions) no longer seemed terribly strange.

8.1   Strings With Space-time Supersymmetry: the Elimination of Tachyons

It was strings with world-sheet supersymmetry which provided the point from which supersymmetry jumped into the realm of field theory and was realized as a space-time symmetry. Later it was an understanding of how string theories can themselves incorporate space-time supersymmetry, that led to the renaissance of strings.

In 1976, Gliozzi, Scherk, and Olive did work that proved to be pivotal in both the study of supersymmetric field theories and in the later study of superstrings.[59,60] As far as the latter is concerned, their work gave a clear indication that a suitable projection of the states of the Ramond-Neveu-Schwarz (RNS) model might respect a space-time supersymmetry. An important characteristic of their projection was that it is free of tachyons. The signal of the latent supersymmetry was that there were an equal number of bose and fermi states at each mass level. Oddly enough, this aspect of their result was not followed up for some time. It seems strange that this important lead was not followed up sooner, especially because in their introduction, the authors put particular emphasis on the possible role of string theories as fundamental theories. They said that "[One] line of thinking is to consider supergravity and its refinements as only a first step.... Then there would be an obvious parallelism between the $V - A$ with coupling constant G_F and the gravitational theory with coupling constant $G_N \sim \kappa^2$ since both constants have the same dimensions. In the renormalizable gauge theories G_F is a phenomenological parameter, expressed as $G_F \sim g^2/M_W^2$ where g is the real dimensionless coupling constant of the theory and M_W^2 plays the role of a cut-off. So one may ask whether one could find a similar expression for $G_N \sim (g^2)^n/\Lambda^2$ and identify g and Λ.... Dual models provide an answer to this kind of question."

The aspect of their work that was immediately followed up was the method that they used to derive supersymmetric field theories. By considering the zero-slope limit while also letting the size of the compact 6-dimensional space go to zero, they derived supersymmetric field theories, some of which were already known and some of which had not yet been constructed. For example a string theory with both open and closed strings had the $N = 4$ supersymmetric Yang-Mills coupled to $N = 4$ supergravity as its field theory limit, while a theory with only close strings yielded $N = 8$ supergravity.

Gliozzi, Scherk, and Olive emphasized the relation to field theory: "...We would like to show that dual models are in accordance with the program of supergravity."[60] This relation was apparently also important to the physicists who took up the challenge of exploring the space-time supersymmetry of the RNS model. In 1982 Green wrote, "The suggestion that a truly supersymmetrical string theory might exist motivated John Schwarz and me to construct the theory explicitly. The ...connection with $N = 4$ Yang-Mills theory and $N = 8$ supergravity provided incentive in addition to the intrinsic interest in string theories as a possible framework for a fundamental theory of nature."[66]

8.2 Explicit Supersymmetry and Anomaly and Infinity Cancelations: Superstrings Take Off

In 1980, approximately four years after the work by Gliozzi, Scherk, and Olive, Green and Schwarz began a systematic program to understand the space-time supersymmetry of 10-dimensional string theory. They described their first paper on the subject as "another step towards a proof of the conjectured supersymmetry of the ten-dimensional theory."[63] In this first paper they actually constructed a supersymmetry generator for the on-shell states and showed that this generator satisfied the supersymmetry algebra and that it also transformed as a Lorentz spinor.

In a series of papers they worked their way toward a covariant description of superstrings. The story of this work, and of the development of a superstring formalism, and of the eventual discovery of anomaly and infinity cancelations, is amply documented in Refs. 115, 69, and in several excellent reviews which are referenced in these. Perhaps the result which caused the most excitement was the discovery that the requirement of anomaly cancelation actually picked out a symmetry group, either $SO(32)$ or $E_8 \times E_8$—this before an $E_8 \times E_8$ superstring theory had been constructed.

The work of Green and Schwarz set off an explosion of further work on superstrings. Although supersymmetry plays an important role in much of this work, it cannot be considered here because of limitations of space. Nevertheless it seems appropriate to at least assess the unusual strength of the trend from a historical perspective. For it does seem that, for a type of theory whose study has not been motivated by experimental observations, string theory has inspired an unprecedented amount of excitement among theorists. (Compare this to the discovery of superconductivity in substances that are maintained at relatively high temperatures. The wave of excitement which has followed this discovery was motivated, and is being constantly fed by, experimental results.)

The fact that something special has indeed happened is documented in the November 9, 1987 issue of *Current Contents*.[56] This issue contains a list of the 1985 physical sciences articles that were most-cited in 1985 and 1986.

In his commentary on the list Eugene Garfield notes the rise of superstrings. "The emergence of superstring theory in this year's study represents an important new development in the field. An indication is the very prominence of the term 'superstring'. In our study of the 1983 papers most-cited in 1983 and 1984, not a single... paper contained the words 'string' or 'superstring' in its title. However by the following year... 'superstrings' appeared in the titles of two papers. This year, 'string' or 'superstring' appear in 24 article titles." In order for a paper to be included on this list of most-cited papers it must have received at least 32 citations. Only 104 papers satisfied this criterion, so the string contingent represents about 23 percent of the total. This is far more than the percentage of papers in any other area. For example there were 12 articles on quasicrystals and 4 on astrophysics. (Papers from all of the "physical sciences" were considered, except for chemistry, which was studied separately.)

Garfield goes on to say that "...this year's study is dominated by papers concerning string theory. And perhaps the most dramatic indication that something important has occurred in the superstring field is that the five top-cited papers are all related and that three have received an unprecedented number of citations for a two-year study in the physical sciences. The most cited paper on this year's list received 304 citations. But more typical of past studies were the 170 citations collected in two years by the paper announcing the discovery of the W^+ and W^- particles."

It seems fitting, that supersymmetry, which (at least in the West) had its origins in the study of string theory, should have had such a profound effect on the study of strings.

9   Prospects for a Superfuture

9.1   Supersymmetry and Experiment

No matter how aesthetically pleasing the concept of supersymmetry is, nor how powerful a tool it provides for the exploration of possible theories, its role as a physical symmetry can be assured only through experimental verification. Verification could come in one of two ways. The first might be experimental discovery of a phenomenon that is predicted uniquely by supersymmetry. An example would be the observation of superpartners of the known particles. Verification could also come in another guise, if the theoretical study of supersymmetry can lead to an understanding of aspects of the physical world which at present must be simply taken as given. If, for example, the masses and coupling constants of the known particles, and perhaps even the dimensionality of space-time and of its fundamental inhabitants could, albeit after the fact, be predicted by supersymmetry, then its value as a physical principle would probably be widely accepted.

A great deal of work has been done to understand the phenomenology of supersymmetry. However there does not seem to be any one phenomenological scenario which would be compelled to occur by the existence of supersymmetry. Therefore, although it might be possible for a series of observations during the next few years to confirm the existence of supersymmetry, there does not seem to be any comparable series of observations which could rule it out. This state of ambiguity is cited by Fayet in his contribution to the 1986 Nobel Symposium.[12] "The observation of superpartners at present or future colliders would probably be considered now as a largely expected discovery. But one should also keep in mind that superpartners might only appear at extremely high energies; or, conversely, that both superpartners and extra dimensions might reveal themselves at future colliders." The situation with regard to possible experimental tests of superstring theories is, if anything, even more ambiguous. Nevertheless work is proceeding to try to determine the observational consequences of superstrings, both for collider experiments and for astrophysics. The motivation for these studies is twofold. Beyond the idea of experimentally testing the validity of the hypothesis of superstring theories, is the question of whether or not (or perhaps which of) these hypotheses are consistent with observations which have already been made.

Given this state of affairs, it is difficult to pick out those themes of the history of the study of supersymmetric phenomenology which will ultimately prove to be most significant. Because of this, and because of the limited space available, the discussion of this theory will be brief. More complete discussions and references can be found elsewhere, for example in Refs. 92, 85, 24, 47.

The earliest phenomenological idea that was related to supersymmetry was the one suggested by Volkov and Akulov, that perhaps one might be able to understand the existence and couplings of the neutrino by viewing it as the Goldstone particle associated with the spontaneous breaking of a fermionic symmetry. The fermionic analogue of the Goldstone theorem was proved by Salam and Strathdee in 1974. This work was referenced by De Wit and Freedman when, in 1975, they studied the phenomenology of Goldstone neutrinos.[49] (They also referenced some unpublished work of W.A. Bardeen.) de Wit and Freedman studied the validity and implications of low energy theorems of spontaneously broken supersymmetry and concluded that, at least for the broad class of theories that they had considered, it was likely that the hypothesis that the neutrino was a Goldstone fermion was inconsistent with experimental results.

The 1974 work of Wess and Zumino illuminated the ability of supersymmetry transformations to connect particles of different spin. In their first paper on supersymmetry they conjectured that "supersymmetry may... provide a natural way for the formulation of higher internal symmetries linking mesons and baryons." After their work, attempts were made to relate the known elementary particles of different spin through supersymmetry transformations. Unfortunately, it soon began to seem unlikely that such a program could be successfully carried out. It is now widely agreed that supersymmetry requires the introduction of so-called superpartners of the known particles. The half integral spin partners of the known bosons are graced with the suffix 'ino'. Thus the graviton has the gravitino, the photon has the photino, while the W's and Z's pal around with the more dubious-sounding winos and zinos. (Before this 'ino' convention was adopted, Bruno Zumino referred to what is now generally called a goldstino as a germion."[116]) The spin zero partners of the leptons and quarks are the sleptons and squarks. Most attempts to detect traces of supersymmetry in the physical world have involved searching for a distinctive signal of the superpartners, and have placed limits on their masses based on the absence of such a signal. (See for example the contribution of Savoy-Navarro to Ref. 92.)

Although it does not seem that supersymmetry can relate all of the fundamental particles to each other within supermultiplets, supersymmetric theories can incorporate multiplets which contain both the spin-0 Higgs bosons and the spin-1 gauge fields. (See for example the contribution of Fayet to Ref. 92.)

Another nice feature of supersymmetric theories is related to the lack of radiative corrections, which was discussed in Sec. V. In the context of supersymmetric grand unified theories, this feature ensures that the widely different mass scales (of the mass of the W and the mass of the X for example) stay widely separated, even when radiative corrections are considered; i.e. hierarchies are preserved in supersymmetric quantum theories.

9.2   A Superfuture?

The flavor of the above discussion reveals that high energy theory and experiment are not coupled in the way that theory and experiment were coupled during the 1920's or even during the 1960's and early 1970's. There are tests of some of the predictions of supersymmetry that either have been done, are ongoing, or are planned. And the continuing interaction between experimentalists and theoreticians is illustrated by the 1982 meeting that took place at CERN, and by the contributions to that meeting which are collected in the *Physics Reports* issue 'Supersymmetry Confronting Experiment'.[92] However, a clear-cut and decisive confrontation between supersymmetric field theory and experiment has yet to occur.

The development of superstrings continues the general trend of a *widening separation between high energy theory and experiment*. The reasons for the trend are not negative. On the contrary, it reflects the successes of our theories to date, which have been capable of enveloping most of the observed high energy phenomena, and our optimism about the ability of the human mind to explore new territory. It is difficult, however, to know where this exploration will take us, or how we will know that we are there. During his talk at the Nobel Symposium, Ellis referred to the "lone experimentalist at this meeting." Glancing through the proceedings of that meeting one also gets a sense of another trend—the increasingly important role that frontier mathematics is playing in the further development of string theory.

Since at any time a new and perhaps unexpected discovery— either experimental or theoretical—could change things dramatically, it is impossible to know definitely what the future holds in store for the further development of supersymmetry. But it is clear that, should a further history of supersymmetry be written ten years from now, it will either include the story of the experimental discovery of supersymmetry, or a more detailed discussion of the ways in which supersymmetry will be known not to manifest itself. A good deal of information of this latter type should be obtainable from observations that are possible within the next decade. *If, at that time, supersymmetry is still a purely theoretical phenomenon, it is possible that it may nevertheless continue to be vigorously studied by theorists. The reason for this is that, beyond its aesthetic appeal, it provides ways of calculating quantities that might otherwise be poorly defined, or at least so much more difficult to compute, that computation is not viable. The lack of experimental verification would, however, present a formidable challenge to theory. For it will probably mean, at least, that either our first attempts at understanding how nature may be supersymmetric have been too naive, or that supersymmetry is a property of the world which can only be understood if one knows about phenomena that occur at much higher energies than those we have thus far been able to observe.*[k]

In considering the possibility that supersymmetry, if it is indeed realized in the physical world, is an ultra high energy phenomenon, we can wonder about how far we can go toward understanding its role through the power of pure thought. One may wonder for example what would have happened if the principle of non-Abelian gauge symmetry had been developed during the last century. Although the Abelian case might have been understood to correspond to electromagnetism, what use would non-Abelian symmetry, beautiful though it is, have been to physicists who were trying to understand the world at much smaller distances than those they were able to probe? Without any knowledge of atomic, much less nuclear structure, and without knowing that the laws of classical physics could not be applied to nature in the small, how much progress would they have been able to make by constructing gauge theories? At this early stage in the development of high energy physics it seems possible that we are in an analogous situation, although we may fervently hope not.

---

[k]We allowed ourselves to italicize this part of the paragraph (written by the author approximately 12 years ago) which turned out to be prophetic. – Editors' note.

Acknowledgments

This paper was written at the Institute for Theoretical Physics at Stony Brook where I have greatly benefited from the presence of several active practitioners of supersymmetry, in particular P. van Nieuwenhuizen. I am also grateful to S. Deser, P. Ramond, A. Salam, D.V. Volkov, and J. Wess for providing helpful information and comments. I would like to thank Max Dresden for telling me about the early conjecture of Bohr, for copies of lecture notes from the 1940's, for his comments on the manuscript, and for his encouragement. I would also like to thank Bruce Gittleman for his help in compiling the data plotted in the graphs, Arvind Borde for useful comments of the manuscript, and Betty Gasparino for typing the list of references.

References

1. E.S. Abers and B.W. Lee, 'Gauge Theories', *Phys. Rep.* C **9**, 1 (1973).
2. Y. Aharonov, A. Casher, and L. Susskind, 'Spin $\frac{1}{2}$ Partons in a Dual Model of Hadrons', *Phys. Rev.* D **5**, 988 (1972).
3. V.P. Akulov and D.V. Volkov, 'Goldstone Fields with Spin 1/2', *Teor. Mat. Fiz.* **18**, 39 (1974).
4. V.P. Akulov and D.V. Volkov, 'Interaction of Goldstone Neutrino with Electromagnetic Field', *ZhETF Pis. Red.* **17**, 367 (1973).
5. V.P. Akulov, D.V. Volkov, and V.A. Soroka, 'Gauge Fields on Superspace with Different Holonomy Groups', *JETP Lett.* **22**, 187 (1975) (*Pis'ma Zh. Eksp. Teor. Fiz.* **22**, 396 (1975); V.P. Akulov, D.V. Volkov, and V.A. Soroka, 'Generally Covariant Theories of gauge Fields on Superspace', *Teor. Mat. Fiz.* **31**, 12 (1977); V.P. Akulov and D.V. Volkov, 'Reimannian Superspace of Minimal Dimensionality', *Teor. Mat. Fiz.* **41**, 147 (1979).
6. T. Appelquist, A. Chodos, P.G.O. Freund, (Eds.), *Modern Kaluza-Klein Theories*, Addison-Wesley, 1987.
7. R. Arnowitt, P. Nath, and B. Zumino, 'Superfield Densities and Action Principle in Curved Superspace', *Phys. Lett.* B **56**, 81

(1975).

8. R. Arnowitt, P. Nath, (Eds.), *Gauge Theories and Modern Field Theory*, MIT Press, 1976.

9. A. Ashtekar, *New Perspectives in Canonical Gravity*, Bibliopolous (Naples) 1988.

10. K. Bardakçi and M.B. Halpern, 'Dual M-Models', *Nucl. Phys.* B **73**, 295 (1974).

11. F.A. Berezin, *The Method of Second Quantization*, Academic Press, 1966.

12. L. Brink, R. Marnelius, J.S. Nilsson, P. Salomonson, and B-S. Skagerstam, Unification of Fundamental Interactions. Proceedings of Nobel Symposium 67, *Physica Scripta* T **15**, (1987).

13. L. Brink, J. Schwarz, and J. Scherk, 'Supersymmetric Yang-Mills Theories', *Nucl. Phys.* B **121**, 77 (1977).

14. L. Brink and P. Howe, 'The $N = 8$ Supergravity in Superspace', *Phys. Lett.* B **88**, 268 (1979).

15. S.M. Christensen, (Ed.), *Quantum Theory of Gravity*, Adam Hilger Ltd., 1984.

16. F. Coester, M. Hammermesh, and W.D. McGlinn, 'Internal Symmetry and Lorentz Invariance', *Phys. Rev.* B **135**, 451 (1964).

17. S. Coleman and J. Mandula, 'All Possible Symmetries of the S-Matrix', *Phys. Rev.* D **159**, 1251 (1967).

18. E. Cremmer and J. Scherk, 'Dual Models in Four Dimensions with Internal Symmetries', *Nucl. Phys.* B **103**, 399 (1975).

19. E. Cremmer and J. Scherk, 'Spontaneous compactification of Space in an Einstein-Yang-Mills-Higgs Model', *Nucl. Phys.* B **108**, 409 (1976).

20. E. Cremmer, B. Julia, and J. Scherk, 'Supergravity Theory in 11 Dimensions', *Phys. Lett.* B **76**, 409 (1978).

21. E. Cremmer and B. Julia, 'The $N = 8$ Supergravity Theory I. The Lagrangian', *Phys. Lett.* B **80**, 48-51 (1978).

22. E. Cremmer, J. Scherk, and J. Schwarz, 'Spontaneously Broken $N = 8$ Supergravity', *Phys. Lett.* B **84**, 83 (1979).

23. R. D'Auria and P. Fré, *Superunification and Extra Dimensions*, World Scientific, 1986.

24. Del Agulia, J.A. de Azcárraga, and L.E. Ibáñez, (Eds.), *Supersymmetry, Supergravity and Related Topics*, World Scientific, 1985.

25. S. Deser and P. van Nieuwenhuizen, 'One-Loop Divergences of Quantized Einstein-Maxwell Fields', *Phys. Rev.* D **10**, 401 (1974).

26. S. Deser and P. van Nieuwenhuizen, 'Nonrenormalizability of the Quantized Einstein-Maxwell System', *Phys. Rev. Lett.* **32**, 245 (1974).

27. S. Deser, H.-S. Tsao, and P. van Nieuwenhuizen, 'Nonrenormalizability of Einstein-Yang-Mills Interactions at the One-Loop Level', *Phys. Lett.* B **50**, 491 (1974).

28. S. Deser and P. van Nieuwenhuizen, 'Nonrenormalizability of Quantized Fermion-Graviton Interactions', *Nuovo Cim. Lett.* **11**, 218 (1974).

29. S. Deser and P. van Nieuwenhuizen, 'Nonrenormalizability of the Quantized Dirac-Einstein System', *Phys. Rev.* D **10**, 411 (1974).

30. S. Deser and B. Zumino, 'Consistent Supergravity', *Phys. Lett.* B **62**, 335 (1976).

31. S. Deser, J. Kay, and K. Stelle, 'Renormalizability Properties of Supergravity', *Phys. Rev. Lett.* **38**, 527 (1977).

32. S. Deser and J. Kay, 'Three-Loop Counterterms for Extended Supergravity', *Phys. Lett.* B **76**, 400 (1987).

33. B. de Wit, P. Fayet, and P. van Nieuwenhuizen, (Eds.), *Supersymmetry and Supergravity '84*, World Scientific, 1984.

34. R. Di Stefano, Ph.D. Dissertation, SUNY at Stony Brook, 1982.

35. R. Di Stefano, 'Disappearance of the Auxiliary Fields in a Canonical Formulation of Supersymmetry', *Phys. Lett.* B **192**, 130 (1987).

36. R. Di Stefano, 'Canonical Construction of Symmetry Generators and Algebras: Application to Simple Supersymmetry (I) Systems without Constraints', ITP:SB:87/81; (II) 'Systems with Constraints', ITP:SB:88/4.

37. D.B. Fairlie and D. Martin, 'New Light on the Neveu-Schwarz Model', *Nuovo Cim.* **18**, 373 (1973).

38. P. Fayet and J. Iliopoulos, 'Spontaneously Broken Supergauge

Symmetries and Goldstone Spinors', *Phys. Lett.* B **51**, 461 (1974).

39. P. Fayet and S. Ferrara, 'Supersymmetry', *Phys. Rep.* **32**, 249 (1977).

40. S. Ferrara and O. Piguet, 'Perturbation Theory and Renormalization of Supersymmetric Yang-Mills Theories', *Nucl. Phys.* B **93**, 261 (1975).

41. S. Ferrara, B. Zumino, and J. Wess, 'Supergauge Multiplets and Superfields', *Phys. Lett.* B **51**, 239 (1974).

42. S. Ferrara and B. Zumino, 'Supergauge Invariant Yang-Mills Theories', *Nucl. Phys.* B **79**, 413 (1974).

43. S. Ferrara and B. Zumino, 'Transformation Properties of the Supercurrent', *Nucl. Phys.* B **87**, 207 (1975).

44. S. Ferrara and P. van Nieuwenhuizen, 'The Auxiliary Fields of Supergravity', *Phys. Lett.* B **74**, 330 (1978).

45. S. Ferrara and J.G. Taylor, (Eds.), *Supergravity '81*, Cambridge University Press, 1982.

46. S. Ferrara, J.G. Taylor, and P. van Nieuwenhuizen, (Eds.), *Supersymmetry and Supergravity '82: Proceedings of the Trieste September 1982 School*, World Scientific, 1983.

47. S. Ferrara, *Supersymmetry*, North-Holland/World Scientific, 1987.

48. D.Z. Freedman and So-Young Pi, 'External Gravitational Interactions in Quantum Field Theory', *Ann. of Phys.* **91**, 442 (1975).

49. D.Z. Freedman and B. de Wit, 'Phenomenology of Goldstone Neutrinos', *Phys. Rev. Lett.* **35**, 827 (1975).

50. D.Z. Freedman and B. de Wit, 'Combined Supersymmetric and Gauge Invariant Field Theories', *Phys. Rev.* D **12**, 2286 (1975).

51. D.Z. Freedman, P. van Nieuwenhuizen, and S. Ferrara, 'Progress Toward a Theory of Supergravity', *Phys. Rev.* D **13**, 3214 (1976).

52. D.Z. Freedman and P. van Nieuwenhuizen, 'Properties of Supergravity Theory', *Phys. Rev.* D **14**, 912 (1976).

53. P.G.O. Freund, Mark A. Rubin, 'Dynamics of Dimensional Reduction', *Phys. Lett.* B **97**, 233 (1980).

54. S. Fubini, d. Gordon, and G. Veneziano, 'A General Treatment of Factorization in Dual Resonance Models', *Phys. Lett.* B **29**, 679 (1969).

55. S. Fubini and G. Veneziano, 'Algebraic Treatment of Subsidiary Conditions in Dual Resonance Models', *Ann. of Phys.* **63**, 12 (1971).

56. E. Garfield, 'The Most Cited 1985 Physical Science Articles', *Current Contents* **27**, 3 (1987).

57. S.J. Gates, M.T. Grisaru, M. Roček, W. Siegel, *Superspace or One Thousand and One Lessons in Supersymmetry*, Benjamin/Cummings, 1983.

58. J.L. Gervais and B. Sakita, 'Field Theory Interpretation of Supergauges in Dual Models', *Nucl. Phys.* B **34**, 632 (1971).

59. F. Gliozzi, J. Scherk, and D. Olive, 'Supergravity and the Spinor Dual Model', *Phys. Lett.* B **65**, 282 (1976).

60. F. Gliozzi, J. Scherk, and D. Olive, 'Supersymmetry, Supergravity Theories and the Dual Spinor Model', *Nucl. Phys.* B **122**, 253 (1977).

61. P. Goddard and C.B. Thorn, 'Compatibility of the Dual Pomeron with Unitarity and the Absence of Ghosts in the Dual Resonance Model', *Phys. Lett.* B **40**, 325 (1972).

62. Y.A. Golfand and E.P. Likhtman, 'Extension of the Algebra of Poincaré Group Generators and Violation of P Invariance', *JETP Lett.* **13**, 323 (1971).

63. M.B. Green and J.H. Schwarz, 'Supersymmetrical Dual String Theory', *Nucl. Phys.* B **181**, 502 (1981).

64. M.B. Green, J.H. Schwarz, and L. Brink, '$N = 4$ Yang-Mills and $N = 8$ Supergravity as Limits of String Theories', *Nucl. Phys.* B **198**, 474 (19821).

65. M.B. Green and J.H. Schwarz, 'Supersymmetrical String Theories', *Phys. Lett.* B **109**, 444 (1982).

66. M.B. Green, 'Supersymmetrical Dual String Theories and Their Field Theory Limits — a Review', Surveys in High Energy Physics **3**, 127 (1983).

67. M.B. Green and J.H. Schwarz, 'Covariant Description of Superstrings', *Phys. Lett.* B **136**, 367 (1984).

68. M.B. Green and J.H. Schwarz, 'Anomaly Cancelations in Su-

persymmetric $D = 10$ Gauge Theory and Superstring Theory', *Phys. Lett.* B **149**, 117 (1984).

69. M.B. Green, J.H. Schwarz, E. Witten, *Superstring Theory*, Cambridge University Press, 1986.

70. M. Grisaru, P. van Nieuwenhuizen, and J.A.M. Vermaseren, 'One-Loop Renormalizability of Supergravity and of the Maxwell-Einstein Theory in Extended Supergravity', *Phys. Rev. Lett.* **37**, 1662 (1976).

71. M. Grisaru, H.N. Pendleton, and P. van Nieuwenhuizen, 'Supergravity and the S-Matrix', *Phys. Rev.* D **15**, 996 (1977).

72. M. Grisaru, 'Two-Loop Renormalizability of Supergravity', *Phys. Lett.* B **66**, 75 (1977).

73. M. Grisaru, W. Siegel, and M. Roček, 'Improved Methods for Supergraphs', *Nucl. Phys.* B **139**, 429 (1979).

74. M. Grisaru, M. Roček, and W. Siegel, 'Zero value for the Three-Loop β function in $N = 4$ Supersymmetric Yang-Mills Theory', *Phys. Rev. Lett.* **45**, 1063 (1980).

75. M. Grisaru and W. Siegel, 'Supergraphity (II). Manifestly Covariant Rules and Higher-Loop Finiteness', *Nucl. Phys.* B **201**, 292 (1982).

76. R. Haag, J.T. Lopuszański, and M. Sohnius, 'All Possible Generators of Supersymmetries of the S-Matrix', *Nucl. Phys.* B **88**, 257 (1975).

77. S.W. Hawking and M. Roček, (Eds.), *Superspace and Supergravity*, Cambridge University Press, 1981.

78. W. Heisenberg, Solvay Ber. Kap. III, IV (1939).

79. P. Howe, K. Stelle, and P. Townsend, 'Miraculous Ultraviolet Cancelations in Supersymmetry Made Manifest', *Nucl. Phys.* B **236**, 125 (1984).

80. J. Iliopoulos and B. Zumino, 'Broken Supergauge Symmetry and Renormalization', *Nucl. Phys.* B **76**, 310 (1974).

81. M. Jacob, (Ed.), *Dual Theory*, North-Holland, 1974.

82. M. Jacob, (Ed.), *Supersymmetry and Supergravity*, North-Holland/World Scientific, 1986.

83. B. Kursonoglu, A. Perlmutter, and L. Scott, (Eds.), *Deeper Pathways in High Energy Physics*, Plenum, 1977.

84. W. Lang and J. Wess, 'Investigation of a Nonrenormaliz-

able Lagrangian Model Invariant Under Supertransformation', *Nucl. Phys.* B **81**, 249 (1974).

85. J. León, J. Pérez-Mercader, and M. Quirós, (Eds.), *Third CSIC Workshop on SUSY and Grand Unification: From Strings to Collider Phenomenology*, World Scientific, 1986.

86. E.P. Likhtman, 'Supergauge Renormalizable Theory of a Massive Vector Field', *JETP Lett.* **21**, 109 (1975).

87. E.P. Likhtman, 'A Supersymmetric Renormalized Theory of a Massive Non-Abelian Vector Field', *JETP Lett.* **22**, 57 (1975).

88. J.T. Lopuszański, 'On Some Properties of Physical Symmetries', *J. Math. Phys.* **12**, 2401 (1971).

89. W.D. McGlinn, 'Problem of Combining Interaction Symmetries and Relativistic Invariance', *Phys. Rev. Lett.* **12**, 467 (1964).

90. B. Milewski, (Ed.), *Supersymmetry and Supergravity 1983*, World Scientific, 1983.

91. C. Montonen, 'Multiloop Amplitudes in Additive Dual-Resonance Models', *Nuovo Cim.* A **19**, 69 (1073).

92. D.V. Nanopoulos and A. Savoy-Navarro, 'Supersymmetry Confronting Experiment', *Phys. Rep.* **105**, 1 (1984).

93. P. Nath and R. Arnowitt, 'Generalized Super-Gauge Symmetry as a New Framework for Unified Gauge Theories', *Phys. Lett.* B **56**, 177 (1975).

94. A. Neveu and J.H. Schwarz, 'Factorizable Dual Models of Pions', *Nucl. Phys.* B **31**, 86 (1971).

95. A. Neveu, J.H. Schwarz, and C.B. Thorn, 'Reformulation of the Dual Pion Model, *Phys. Lett.* B **35**, 529 (1971).

96. A. Neveu and J. Scherk, 'Connection Between Yang-Mills Fields and Dual Models', *Nucl. Phys.* B **336**, 155 (1972).

97. M. Nouri-Moghadam and J.G. Taylor, 'One-Loop Divergences for the Einstein-charged meson system', *Proc. Roy. Soc.* A **344**, 87 (1975).

98. L. O'Raifeartaigh 'Internal Symmetry and Lorentz Invariance', *Phys. Rev. Lett.* **14**, 332 (1965).

99. L. O'Raifeartaigh 'Mass Differences and Lie Algebras of Finite Order', *Phys. Rev. Lett.* **14**, 575 (1965).

100. L. O'Raifeartaigh 'Lorentz Invariance and Internal Symmetry', *Phys. Rev.* B **139**, 1052 (1964).

101. A. Pais, *Inward Bound: of Matter and Forces in the Physical World*, Oxford University Press, 1986.
102. K. Pilch, P. Townsend, and P. van Nieuwenhuizen, 'Compactification of $d = 11$ Supergravity on S^4 (or 11=7+4, too)', *Nucl. Phys.* B **242**, 377 (1984).
103. P. Ramond, 'Dual Theory for Free Fermions', *Phys. Rev.* D **3**, 2415 (1971).
104. R. Ruffini, (Ed.), *Proceedings of the First Marcel Grossman Meeting on General Relativity*, North Holland, 1975.
105. A. Salam and J. Strathdee, 'Super-Gauge Transformations, *Nucl. Phys.* B **76**, 477 (1974).
106. A. Salam and J. Strathdee, 'Super-Symmetry and Non-Abelian Gauges', *Phys. Lett.* B **51**, 353 (1974).
107. A. Salam and J. Strathdee, 'SU(6) and Supersymmetry', *Nucl. Phys.* B **84**, 127 (1975).
108. A. Salam and J. Strathdee, 'Superfields and Bose-Fermi Symmetry', *Phys. Rev.* D **11**, 1521 (1975).
109. A. Salam and J. Strathdee, 'Feynman Rules for Superfield', *Nucl. Phys.* B **86**, 142 (1975).
110. J. Scherk, 'Zero-Slope Limit of the Dual Resonance Model', *Nucl. Phys.* B **31**, 222 (1971).
111. J. Scherk and J.H. Schwarz, 'Dual Models for Non-Hadrons', *Nucl. Phys.* B **81**, 118 (1974).
112. J. Scherk and J.H. Schwarz, 'Dual Field Theory of Quarks and Gluons', *Phys. Lett.* B **57**, 463 (1975).
113. J. Scherk and J.H. Schwarz, 'Spontaneous Breaking of Supersymmetry Through Dimensional Reduction', *Phys. Lett.* B **82**, 60 (1979); *Nucl. Phys.* B **153**, 60 (1979).
114. J.H. Schwarz, 'Dual Resonance Theory', *Phys. Rep.* **8C**, 269 (1973).
115. J.H. Schwarz, (Ed.), *Superstrings: the First Fifteen Years*, World Scientific, 1985.
116. J.R. Smith, ed.: *Proceeding of the XVII International Conference on High Energy Physics*, Science Research Council, Rutherford Laboratory, 1974.
117. M. Sohnius and P. West, 'Conformal Invariance in $N = 4$ Supersymmetry Yang-Mills Theory', *Phys. Lett.* B **100**, 241

(1981).

118. K. Stelle and P.C. West, 'Minimal Auxiliary Fields for Supergravity', *Phys. Lett.* B **74**, 333 (1978).

119. G.'t Hooft and M. Veltman, 'One-Loop Divergences in the Theory of Gravitation', *Ann. Inst. Henri Poincaré* **20**, 69 (1974).

120. P. van Nieuwenhuizen, M.T. Grisaru, and J.A.M. Vermaseren, 'One-Loop Renormalizability of Pure Supergravity and the Maxwell-Einstein Theory in Extended Supergravity', *Phys. Rev. Lett.* **37**, 1662 (1976).

121. P. van Nieuwenhuizen and P.K. Townsend, 'Geometrical Interpretation of Extended Supergravity', *Phys. Lett.* B **67**, 439 (1977).

122. P. van Nieuwenhuizen and D.Z. Freedman, *Supergravity: Proceedings of the Supergravity Workshop at Stony Brook*, North-Holland, 1979.

123. P. van Nieuwenhuizen, 'Supergravity', *Phys. Rep.* **68**, 189 (1981).

124. D.V. Volkov, 'Phenomenological Lagrangians', *Sov. J. Particles Nucl.* **4**, 1 (1973).

125. D.V. Volkov and V.P. Akulov, 'Possible Universal Neutrino Interaction', *JETP Lett.* **16**, 438 (1972).

126. D.V. Volkov and V.P. Akulov, 'Is the Neutrino a Goldstone Particle?', *Phys. Lett.* B **46**, 109 (1973).

127. D.V. Volkov and V.A. Soroka, 'Higgs Effect for Goldstone Particles with Spin $\frac{1}{2}$', *Pis'ma Zh. Eksp. Teor. Fiz.* **18**, 529 (1973); D.V. Volkov and V.P. Akulov, 'Is the Neutrino a Goldstone Particle?', *Phys. Lett.* B **46**, 109 (1973); D.V. Volkov and V.A. Soroka, 'Gauge Fields for Symmetry Group with Spinor Parameters', *Teor. Mat. Fiz.* **20**, 291 (1974).

128. J. Wess and B. Zumino, 'Supergauge Transformations in Four Dimensions', *Nucl. Phys.* B **70**, 39 (1974).

129. J. Wess and B. Zumino, 'A Lagrangian Model Invariant under Supergauge Transformations', *Phys. Lett.* B **49**, 52 (1974).

130. J. Wess and B. Zumino, 'A Supergauge Invariant Extension of Quantum Electrodynamics', *Nucl. Phys.* B **78**, 1 (1974).

131. J. Wess and J. Bagger, *Supersymmetry and Supergravity*, Princeton University Press, 1983.

132. P. West, *Introduction to Supersymmetry and Supergravity*, World Scientific, 1986.
133. C. Wetterich, 'Dimensional Reduction of Weyl, Majorana and Majorana-Weyl Spinors', *Nucl. Phys.* B **222**, 20 (1983).
134. E. Witten, 'Search for a Realistic Kaluza-Klein Theory', *Nucl. Phys.* B **186**, 412 (1981).
135. C.N. Yang and R.L. Mills, 'Conservation of Isotopic Spin and Isotopic Gauge Invariance', *Phys. Rev.* **96**, 191 (1954).
136. T. Yoneya, 'Quantum Gravity and the Zero-Slope Limit of the Generalized Virasoro Model', *Nuovo Cim. Lett.* **8**, 951 (1973).
137. T. Yoneya, 'Connection of Dual Models to Electrodynamics and Gravidynamics', *Prog. Theor. Phys.* **51**, 1907 (1974).
138. B. Zumino, 'Supersymmetry and the Vacuum', *Nucl. Phys.* B **89**, 535 (1975).
139. W. Nahm, 'Supersymmetries and Their Representations', *Nucl. Phys.* B **135**, 149 (1978).

EPILOGUE

It is with deep sorrow that we have learned of the death of Professor Lochlainn O'Raifeartaigh on 18 November 2000. Just a few weeks before, he gave an engaging talk at the Symposium "30 Years of Supersymmetry" organized by the Theoretical Physics Institute in Minnesota. He was enthusiastic about supersymmetry on lattices, and outlined a program the final goal of which was solving supersymmetric gauge theories at strong coupling.

Prof. O'Raifeartaigh held closely to his heart the idea of this Volume. He presented several successive versions of his contribution (see p. 145) and polished them until he was fully satisfied. Lochlainn O'Raifeartaigh's work on mechanisms of supersymmetry breaking will forever be a part of the history of the development of supersymmetric theories.

The Editors